Permaculture Company

Beyond Hustle, How to Build a Regenerative Company That Lasts

Rich Peterson

Generous Books

Contents

Start Here

From Extraction to Regeneration: A Call to the Builders of the Next Economy

If you're reading this book, you probably sense it too.
Something in our current way of building companies — of measuring success, of defining growth, feels spent.

We've optimised everything except what truly matters.
The planet is exhausted.
People are burned out.
And yet the creative impulse, the desire to build something good, lasting and alive, remains.

This book is a bridge between worlds.
It invites founders, makers and stewards to look to nature not as a metaphor, but as a manual.

For all of time, living systems have faced every challenge a company will ever face:
growth, competition, scarcity, succession, death and renewal.
Nature solved them all through design, not domination.

You'll find no hustle hacks here.
No promises of viral growth or 10x returns.
Instead, you'll find something older and wiser: the patterns of life itself, distilled into practical principles you can apply to your team, your product, your customers and yourself.

If you've ever wondered whether it's possible to run a business that feels alive instead of draining...
The forest already answered that question.

Your task now is to listen.

Why This Book Exists

I wrote this book because I kept meeting founders who were quietly dying inside.
They loved their work. They loved their people. But they were trapped in a system that rewarded depletion over depth, growth over groundedness.

They didn't want to quit capitalism. They wanted to redesign it, from the inside out.

Permaculture offered a blueprint: a design system for life that regenerates what it touches.
When applied to gardens, it restores soil.
When applied to communities, it restores trust.
And when applied to companies, it restores meaning and energy.

This book is my attempt to translate those ecological patterns into business practice.
Not as theory, but as design language, practical, visual and alive.

Each chapter offers a pattern you can use immediately.
Each field guide helps you apply it.
And together, they form a regenerative operating system that will change how you see business forever.

You don't need to be a farmer, activist, or academic to use this.
You just need to care about creating systems that last, and people who thrive.

If you're burned out by hustle but not ready to give up on building, this book is for you.

How to Use This Book

A Field Guide for Builders, Not Spectators

This is not a book to read once and shelve.
It's a toolkit. A lens. A set of living instructions.

You can use it three ways:

1. **Sequentially, as a journey.**
   Each chapter builds on the last, guiding you from ethics (roots) to principles (branches) to systems (canopy).
   Ideal for founders designing or redesigning from scratch.

2. **Situationally, as a reference.**
   Each principle is self-contained. Open to any chapter that matches your current challenge.
   For example:

   - Facing burnout? → Chapter 6, *Use and Value Re-*

newable Resources.

- Struggling with silos? → Chapter 9, *Integrate Rather Than Segregate.*

- Feeling chaotic growth? → Chapter 10, *Use Small and Slow Solutions.*

3. **Practically, as a workbook.**
Each chapter ends with a *Field Guide*, a short set of exercises you can implement immediately.
They're designed to fit into your current rhythm, not add more to your plate.

The principles are timeless, but your application will evolve.
Use this book like a compass, not a map.
It points toward direction, not destination.

Return to it when things feel unclear, when you need to remember how living systems work.
Because your company is one.

Chapter 1. Care for the Earth, Care for People, Fair Share

The Ethics of Living Business

In a forest, no species grows at the expense of the whole.

The tree takes nutrients from the soil, but its leaves return them. The mycelium feeds on decay, but in doing so, makes new life possible. The system works because every part obeys an unspoken code: take only what you need, share what you can and give back more than you consume.

This is nature's ethics.
And it's the only reason life still exists.

The Three Ethics of Regeneration

When Bill Mollison and David Holmgren first articulated the ethics of permaculture, **Care for the Earth, Care for People and Fair Share**, they weren't writing philosophy. They were describing how living systems sustain themselves over time.

Each ethic corresponds to a design rule:

- **Care for the Earth**, Maintain the conditions for life.
- **Care for People**, Ensure energy circulates, not concentrates.
- **Fair Share**, Distribute surplus to strengthen the whole.

Translated to business, these become the foundation for a **regenerative operating culture**:

- *Steward resources rather than extract them.*
- *Design work systems that sustain human energy.*
- *Treat profit as yield, a resource to reinvest, redistribute, or renew.*

Let's make this practical.

1. Care for the Earth → Stewardship of Resources

In nature, "care for the earth" means maintaining the fertility of soil, the purity of water and the cycles of regeneration.

In business, it means **designing flows that renew rather than consume**.

Most founders interpret this as "sustainability." But sustainability tends to stop at compliance, recycling bins, carbon offsets, or mission statements. Stewardship runs deeper. It means treating every resource, material, digital, human, as a **renewable system**, not a consumable.

Practical applications:

1. **Design closed loops.** Reuse byproducts within your own processes. Interface Carpets, for instance, recycles nylon scraps into new flooring, cutting virgin material use by over 60%.

2. **Track energy, not just money.** Patagonia measures resource impact per product line, not to boast, but to design lighter patterns. You can do this too: calculate the "energy intensity" of your core offerings, energy per sale, per unit, or per customer.

3. **Align suppliers with values.** Ask: who grows our materials, codes our systems, delivers our goods? Are we regenerating or depleting those ecosystems?

Even in a service business, stewardship applies.
A marketing agency that chooses long-term retainers over endless churn is conserving energy, creating fewer but deeper relationships.

Stewardship doesn't slow you down. It keeps your soil fertile.

2. Care for People → Designing for Human Energy

In forests, health emerges from interdependence. A tree's shade protects seedlings. Fungi share nutrients across species lines. Diversity ensures resilience.

In companies, the same rule applies: **if people are depleted, the ecosystem collapses.**

Most organisations talk about "people-first" values, but then design work systems that treat human attention as an infinite input.
Regenerative founders reverse that assumption: they design for **energy cycles** instead of time blocks.

In practice, that means:

- **Energy-based work design.** Organise work around natural rhythms: creative tasks in the morning, meetings in the afternoon, rest embedded in the week.
- **Autonomy as a nutrient.** Basecamp calls this "calm work." No status meetings, few real-time chats and six-week cycles with rest intervals.
- **Shared ownership models.** Zingerman's Community of Businesses distributes equity to employees, turning labor into stewardship.

Observation exercise:
Map your team's energy over a typical week. Where do people light up? Where do they flatten out? Those peaks and valleys

are your company's "photosynthesis map." Use it to redesign meeting schedules, creative cycles, or roles.

Care for people is not a perk. It's the foundation of productivity that lasts.

3. Fair Share → Yield as Circulation

Forests don't hoard. When one tree drops fruit, others feed. Surplus is never stored for status; it's distributed for stability.

In regenerative business, **profit is not extraction, it's yield**. And yield only matters when it circulates.

Three ways to practice Fair Share:

1. **Reinvest locally.** Zingerman's reinvests profits into staff-owned spin-offs, bakeries, roasters, training programs, creating a resilient local network rather than a fragile empire.

2. **Transparent surplus sharing.** Buffer, a digital company, publishes salaries and revenue data openly; when profits rise, bonuses distribute proportionally to contribution.

3. **Circular value design.** Dr. Bronner's invests a portion of profits into regenerative agriculture projects that rebuild the supply chain it depends on, from coconut farms in Sri Lanka to palm groves in Ghana.

Fair Share doesn't mean charity. It's ecological accounting: returning surplus to strengthen the system that generated it.

Ask yourself:
Where does your company's surplus go, back into health or out into extraction? The answer will tell you whether your business is an ecosystem or an engine.

Ethics as Operating Constraints

Ethics aren't slogans for the About page; they're **design constraints** that guide decisions under pressure.

- **When growth tempts you to cut corners, "Care for the Earth"** reminds you to protect long-term fertility.
- **When efficiency demands more from exhausted teams, "Care for People"** brings you back to rhythm.
- **When profit rises faster than purpose, "Fair Share"** redistributes yield before imbalance sets in.

These three checks form a simple but powerful internal compass. They turn values into parameters that shape every system you build, hiring, pricing, operations and governance.

Example:
A regenerative café owner sets a rule: staff can't work more than 32 hours per week without profit-sharing. That single policy aligns all three ethics, it sustains people, limits extraction and shares yield.

Another founder institutes "material caps": no product can use more resource energy than it returns in customer lifetime value. It sounds restrictive, but in practice, it fuels innovation, forcing designers to create longer-lasting, higher-margin offerings.

Ethics become boundaries that generate creativity.

The Mycorrhizal Metaphor: Hidden Networks of Care

Beneath every forest lies a web of fungal threads, the **mycorrhizal network**, connecting roots, transferring nutrients, sending distress signals. When one tree suffers drought, others share water. When one thrives, it feeds the rest.

Healthy organisations have invisible networks too. They're the relationships, trust bonds and informal support systems that keep everything alive when formal structures fail.

In regenerative companies, leaders tend these networks deliberately:

- They host **circles of reflection** where people share what's draining or feeding them.
- They design **peer mentorship** rather than top-down management.
- They invest in **cultural soil**, rituals, shared meals, silence, as seriously as product development.

The health of your invisible networks determines how your company weathers storms.
Feed them first.

Case in Point: Patagonia's Ethical Flywheel

Patagonia embodies the triple ethic as an operational system:

- **Care for the Earth:** Repair programs, lifetime warranties and material transparency.
- **Care for People:** Employee activism leave, child care at HQ and equitable profit sharing.
- **Fair Share:** Donating 1% of sales and eventually transferring ownership to a climate trust.

Their decisions are not reactive CSR but **ethical design loops**. Each choice reinforces the others, better products lead to higher margins, which fund environmental repair, which deepens brand trust, which stabilises revenue.

That's the regenerative flywheel: **Ethics → Trust → Value → Surplus → Regeneration → Trust.**

You can build this too, at any scale.
It starts with ethics as design inputs, not marketing outputs.

Field Guide: Applying the Three Ethics

Use this as a diagnostic and design prompt for your own company.

1. Care for the Earth, Resource Stewardship

- List your top five material or energetic inputs. For each, ask: *Is this Renewable? Reusable? Repairable?*
- Identify one place you can turn waste into a resource this quarter. Example: Convert unused content into an evergreen library; repurpose offcuts into prototypes.

- Commit to a "fertility metric", a way to track how your operations restore, not deplete (energy per outcome, waste diverted, or hours restored).

2. Care for People, Energy Health

- Map a week in your organisation's energy flow. When are people most alive? When are they most drained?
- Create a "Rest Loop": predictable cycles of focus, completion and renewal.
- Redesign one meeting, workflow, or policy to reduce chronic depletion.

3. Fair Share, Yield Circulation

- Define your company's "surplus." (It might be money, time, or knowledge.)
- Decide how much of that surplus should be reinvested, redistributed, or gifted to strengthen the whole system.
- Create a visible mechanism for this flow, a profit-sharing pool, open-source toolkit, or community partnership.

Then, write your company's version of the three ethics in plain language, short enough to fit on a whiteboard, strong enough to guide every decision.

Takeaway

Ethics aren't a moral accessory; they're the soil structure of a regenerative company.
When you design around **Care for the Earth, Care for People and Fair Share**, your business begins to act less like a factory and more like a forest: rooted, adaptive, generous and alive.

In the chapters that follow, we'll move from ethics to **principles**, from why regeneration matters to how you build it day by day.

Next up:
Chapter 2. Observe and Interact, See Before You Scale.

Chapter 2. Observe and Interact, See Before You Scale

The First Skill of Regenerative Founders

In nature, nothing rushes to grow.

The forest doesn't start with planting. It starts with noticing. The permaculturist begins by sitting still, watching how sunlight moves, how water pools, where wind scours the soil, which species thrive without intervention.

Only after seasons of observation does design begin.
The principle is simple: **work with what's alive, not what's assumed.**

In business, the same rule applies.
Before scaling systems, products, or teams, regenerative founders **observe the ecosystem they already have.**

Because what you don't see, you can't sustain.

The Ecologist vs. The Engineer

Most entrepreneurs are trained as engineers. They diagnose a problem, design a solution — then deploy it as fast as possible. Success is measured in output: lines of code, campaigns shipped, products launched.

Regenerative founders act more like ecologists. They observe patterns first, then intervene gently, test and learn from feedback. The difference is subtle but transformative.

The Engineer	The Ecologist
Plans from abstraction	Designs from observation
Seeks efficiency	Seeks balance
Measures output	Measures flow
Controls systems	Tends relationships

The engineer optimises for predictability; the ecologist designs for resilience.
Both are useful. But in living systems, and living companies, **resilience beats efficiency over time.**

Reading Your Organisational Landscape

Observation in business isn't passive. It's a structured practice, a way of seeing the hidden dynamics that shape performance and culture.

Here's how ecological observation translates into company design:

1. Watch the Energy Flows

In nature, energy is sunlight, water, nutrients. In business, energy is attention, trust, money and morale.
Observe where these are flowing, and where they're leaking.

Try this:

- Trace a recent project from start to finish. Who carried most of the weight? Where did communication or motivation break down?

- Follow your cash flow the same way: which activities generate the most stability per unit of effort?

Patterns will emerge: a few processes that generate energy and a few that consume it.
Your job is to stop patching leaks with hustle and start redesigning the system to flow naturally.

2. Map the Microclimates

Every ecosystem has microclimates, warm corners, shaded hollows, frost pockets.
Organisations do too. Teams, products, or clients create different conditions for growth.

Some are fertile (trust, clarity, aligned incentives). Others are barren (resentment, overload, confusion).
You can't treat them all the same.

Practical step:

- Create a "company climate map." Rate each area (department, client group, product line) on two scales: **Energy Health** and **Yield.**
- Ask: where does effort feel easy and meaningful? Where does it feel forced or draining?

You'll discover where your company's sunlight and shade truly lie.
Don't rush to fix. First, see.

3. Notice the Edge Zones

In ecosystems, the edges, where forest meets field or river meets soil, are the most fertile.
New life happens in transition zones.

In companies, edges are the boundaries between teams, roles, or disciplines. They're where innovation and friction coexist.

Example:
At IDEO, project teams deliberately combine designers, engineers and anthropologists. The tension between perspectives creates new insights, the "edge effect."

Implementation tip:

- During meetings, observe who speaks at the edges. Are engineers listening to marketing? Are finance and operations sharing data?
- Don't eliminate friction. Frame it. Design conversations to let diversity pollinate ideas.

Healthy edges produce creativity; ignored edges breed silos.

Case Study: Basecamp, Learning from the Landscape

When Basecamp (then 37signals) built its first project management app, they didn't start by coding.
They spent months **observing clients' pain points** in design projects, email chains, endless meetings, chaotic deadlines.

Instead of assuming what users needed, they asked, "Where does the real energy go in a project?"
Their insight: projects fail not from lack of effort but from unclear communication.

Basecamp's software was a direct response to that observed pattern: simplicity, asynchronous updates, calm collaboration.
Twenty years later, it remains profitable, stable and small by design, precisely because they **observe before they scale.**

As Jason Fried put it, "We didn't design the product we wanted to sell; we designed the solution we wanted to use."

Observation as Leadership Practice

For founders, observation is not just analysis, it's culture.
It starts with how you pay attention to people.

Here's how regenerative leaders observe:

1. **Without judgment.** They listen to patterns, not complaints.

2. **With curiosity.** They ask "What's trying to happen

here?" instead of "Who's at fault?"

3. **Through multiple lenses.** They see through finance, operations, emotion and ecology simultaneously.

Observation creates psychological safety. When people feel seen, they reveal what's really going on, and that data is pure gold for redesign.

Practical Tool: The Organisational Ecosystem Map

Take a blank sheet of paper.
Draw your company as if it were a landscape.

1. **Centre:** The core, you, your leadership team, or your central product.
2. **Surroundings:** Key stakeholders, staff, clients, suppliers, community.
3. **Flows:** Arrows showing where energy (time, money, trust) moves.
4. **Hotspots:** Circles where energy accumulates (overload) or drains (bottlenecks).
5. **Edges:** Points where departments or systems intersect.

Now step back and ask:

- What's thriving with little input?
- What's draining energy faster than it replenishes?

- What could you stop doing and the system would heal itself?

This single map will teach you more about your company than a dozen strategy offsites.

The Danger of Scaling Blind

Scaling without observation is like planting a forest on asphalt. You can pour all the money and motivation you want, nothing will take root.

Startups often scale what's working in numbers, not in patterns. They multiply output but also multiply dysfunction.
Without observation, growth amplifies imbalance.

Example:
A purpose-driven agency grows fast on project revenue but never tracks the emotional cost of client churn. Burnout hits. The founder blames workload, not realising the true leak was relational trust.

Observation surfaces the invisible, the real causes, not the symptoms.

Observation Systems, Not Just Habits

To make observation a repeatable part of your company, design **systems for noticing.**

1. Reflection Rhythms

Build in regular pauses to look at what's actually happening.

- **Weekly retrospectives:** "What energised us? What drained us?"
- **Monthly pattern reviews:** Collect one-page "field notes" from each team.
- **Quarterly system scans:** Revisit your ecosystem map.

2. Decision Logs

Every major decision gets logged with its "why." Review these quarterly to see if the reasons still hold.
You'll begin spotting recurring blind spots, your organisational weather patterns.

3. Stakeholder Listening

Like soil testing, regular listening reveals hidden depletion. Survey customers, staff and partners, not just for satisfaction, but for **energy health.** Ask: "What feels sustainable? What feels strained?"

These small feedback loops turn observation from intuition into data.

Case in Point: The Bee's Eye View

Bees don't manage the hive from above. They sense locally. Each bee responds to micro-conditions, temperature, nectar flow, scent. The collective intelligence emerges from millions of small acts of observation.

Your company can do the same.
Encourage local noticing: frontline employees, customer support staff and junior designers often see energy shifts long before leadership does.

Practical idea:

- Launch a "Signals Channel", a shared digital space for micro-observations: small patterns noticed in work or customer behaviour.
- Rotate the role of "observer of the week" to summarise patterns, not problems.

Over time, you'll build distributed awareness, the foundation of adaptability.

Observation → Interaction

Observation only becomes useful when paired with gentle action.
In permaculture, this means introducing small, reversible interventions, pruning, redirecting water, adding compost, and then watching how the system responds.

In business, the same applies: **small experiments before big bets.**

- Pilot a new workflow with one team.
- Test a shorter week for a month.
- Reallocate one budget line from marketing to community building.

Then, observe the results not just in metrics, but in energy. Did tension ease? Did creativity increase? Did trust deepen?

That's living feedback, the lifeblood of regenerative design.

Field Guide: Practising Observation in Your Company

1. Weekly "Sit Spot"

Borrowed from ecology, a "sit spot" is a fixed place you return to, to notice change over time. In business, your sit spot might be a recurring meeting, Slack channel, or client relationship.
Pick one. Commit to observing it weekly for a month.
Ask: What's shifting? What repeats? What wants to emerge?

2. The 80/20 Energy Audit

List your top 10 activities, projects, or clients.
Mark each as **Energising (+)** or **Draining (–)**.
Then apply the 80/20 rule: double down on the top 20% of energising work and design systems to phase out or redesign the bottom 20% draining work.

3. The Edge Map

Draw your organisational edges, places where roles, teams, or markets overlap.
Edges might be "Sales & Marketing," "Product & Customer Service," or "Finance & Impact."
Label which edges are fertile (new ideas, collaboration) and

which are eroding (misalignment, duplication).
Decide one small experiment to activate a fertile edge this quarter.

4. Reflection Circle

Once per month, host a short, non-judgmental meeting:

- What gave you energy this month?
- What drained it?
- What did you notice that we're ignoring?

No solutions allowed. Just noticing.
Over time, patterns emerge that point to next steps.

Takeaway

Observation is not passive; it's the first act of intelligent design.

When you observe before you scale, you create alignment between energy and effort.
You stop forcing growth and start following it.

The regenerative founder doesn't chase momentum; they **cultivate awareness.**

Everything healthy in nature begins with observation, so does every company that lasts.

Up Next

Chapter 3. Catch and Store Energy, Build Lasting Assets. We'll move from seeing your ecosystem to strengthening it, learning how to capture and store value before it evaporates.

Chapter 3, Catch and Store Energy: Build Lasting Assets

Designing Businesses That Retain What They Create

The Lesson of the Rain Barrel

In nature, rain is generous but unpredictable. A healthy ecosystem doesn't just rely on rainfall, it captures it.
The soil stores water in its pores. Trees store it in their roots. Ponds and aquifers hold it in reserve for dry seasons.

That simple act, **catching and storing energy before it evaporates**, is what allows life to persist.

In business, most founders do the opposite. They work hard to generate energy, attention, money, trust, creativity, but let it drain away as fast as it arrives.

A viral marketing campaign disappears after a week.
A brilliant employee leaves, taking their knowledge with them.
A project succeeds, but no process is documented.

Each leak costs more than money; it costs momentum.

The principle is simple:
If you don't store your energy, you'll always need to create more.

What Counts as Energy in a Company

Before you can store energy, you have to know what it is.

In a business ecosystem, energy takes multiple forms:

Type of Energy	Description	Storage Form
Financial Energy	Cash flow, revenue stability, reserves	Profit buffers, recurring income
Human Energy	Motivation, creativity, relationships	Rest cycles, documentation, culture
Social Energy	Trust, reputation, partnerships	Brand equity, testimonials, networks
Intellectual Energy	Knowledge, systems, ideas	Playbooks, templates, internal wikis
Creative Energy	Innovation, design insight	Prototyping loops, shared libraries

Regenerative founders learn to **harvest each form intentionally.**
They design systems that capture value and make it reusable, like barrels for rain.

The Problem of Leaky Systems

Most businesses are sieves disguised as engines.

- They attract customers but fail to capture loyalty.
- They train employees but lose institutional memory when people leave.
- They produce content but never reuse it.

Every time energy exits without being stored, the founder must compensate through hustle, more effort, more output, more stress.

That's why so many small businesses plateau around the same stage.
They mistake motion for progress.

The regenerative alternative is to **turn motion into momentum**, by storing the energy you already generate.

Case Example: Kit, Building Creative Capital

Kit, an email platform for creators, realised early that its energy wasn't in technology, it was in community.
Instead of competing on features, they invested in **relationships with creators**: sponsoring small conferences, spotlighting customer stories and sharing transparent revenue data.

Each connection became stored social energy, trust that compounded over time.

By the time competitors entered the market with cheaper tools, ConvertKit's "creative capital" made it nearly immune to churn. Customers stayed because they felt seen.

They didn't just capture users. They stored belonging.

From Cash Flow to Energy Flow

Most founders think "store energy" means "build savings." That's part of it, but the deeper shift is **thinking in flows, not balances.**

Just as a forest's soil must stay moist but not flooded, your company's cash, attention and morale must circulate without leaking or stagnating.

Financial Energy

- **Build yield buffers.** Save at least one "season's worth" of operating expenses. It's not caution; it's compost, the nutrients that let you innovate without panic.

- **Shift from one-time sales to recurring relationships.** Subscriptions, retainers, or membership programs mimic perennial plants, steady harvests instead of seasonal surges.

- **Design circular spending.** Partner with suppliers and customers who feed your ecosystem. Money that stays in your loop multiplies value.

Human Energy

- **Capture learning.** Every project should end with a short debrief: what worked, what didn't, what we'll reuse. Store this in a visible place, your "organisational soil."
- **Codify culture.** Write down not just processes, but principles: how you decide, how you rest, how you resolve conflict. Culture leaks fastest when it's unwritten.
- **Build regenerative rhythms.** Rotate roles, allow sabbaticals, design rest weeks. Energy stored in people's health pays the highest dividends.

Case Example: Buffer, Storing Transparency as Trust

Buffer didn't invent social media management, they made it **transparent.**
From day one, every salary, profit figure and experiment was shared publicly.

That radical openness created stored social energy: trust, goodwill and accountability.

When revenue dipped, customers rallied. When hiring slowed, their audience sent referrals.
Buffer's transparency became a renewable energy source, self-sustaining trust.

The lesson:
Store what others waste, honesty, community, integrity, and it becomes a competitive advantage.

Building Energy Reservoirs

Let's bring this down to practice.
Regenerative companies intentionally create **energy reservoirs**, structures that collect, concentrate and reuse value.

1. Knowledge Reservoirs

Create living documentation:

- **"After Action" journals** summarise each major project.
- **Pattern libraries** for recurring decisions (pricing, hiring, communication).
- **Shared databases** of lessons learned.

When new people join, they draw from this reservoir instead of draining yours.

2. Relationship Reservoirs

Don't let networks evaporate after one transaction.
Design **relationship loops**:

- Reconnect with past clients quarterly, not to sell but to share value.

- Create a private community for your most engaged customers.
- Celebrate alumni employees, keep the ecosystem open.

Relationships are your watershed; tend them before they run dry.

3. Creative Reservoirs

Store your creative breakthroughs.

- Keep a **prototype library**, old drafts, designs, campaign ideas.
- Archive stories and images that capture your brand's spirit.
- Create "creative compost" sessions: revisit old ideas to see what's sprouting anew.

Even failed ideas decompose into fertile ground.

Pattern: Capture Before Completion

In ecosystems, the moment of yield is also the moment of greatest loss, fruit falls, nutrients leach.
So nature builds redundancies: seeds scatter, roots absorb, fungi feed.

In companies, every completion moment should include a **capture moment.**

- Before closing a project, harvest learnings.

- Before launching a new product, record your decision logic.
- Before someone leaves the team, host an "exit harvest" conversation.

You'll discover that most of what you thought was fleeting, experience, insight, goodwill, can be stored with simple rituals.

Energy Storage as Equity Design

Equity isn't just ownership; it's storage.
When you share ownership with employees or partners, you store human and social energy inside the company rather than letting it drain into turnover.

Zingerman's exemplifies this. Each spin-off business is co-owned by those who run it.
The energy of entrepreneurship, creativity, pride, responsibility, is captured in equity.

That's why Zingerman's can expand without collapsing. Each new venture strengthens, rather than depletes, the whole.

The Mycelial Model of Asset Building

In forests, the mycelial network stores and redistributes nutrients based on need. It's decentralised and adaptive.

Regenerative companies build similar **distributed asset systems.**
Instead of centralising everything (money, knowledge, decision-making), they spread capacity across the ecosystem.

- Shared resource libraries instead of private silos.
- Team-level budgets with autonomy.
- Peer mentorship replacing top-down management.

This ensures that stored energy remains alive, constantly moving, never hoarded.

Stored energy that stagnates becomes bureaucracy; stored energy that flows becomes resilience.

Field Guide: Catch and Store Energy in Practice

1. Energy Leak Audit

List the top five energy forms in your business:

- Financial
- Human
- Social
- Intellectual
- Creative

For each, ask:

- Where is this energy generated?
- Where is it leaking?
- What small structure could store it? (A policy, tool,

ritual, or document)

Act on one leak per quarter. Slow improvement compounds.

2. Create a Knowledge Reservoir

Start an "Ecosystem Log", a shared document everyone can access.
After each project:

- Record what worked, what didn't and what patterns emerged.
- Tag learnings so they can be found later.

This becomes your organisation's living compost heap, every failure decomposes into future growth.

3. Design a Buffer Zone

In ecosystems, buffers protect against drought.
In business, buffers protect against burnout and panic.

- Financial buffer: one quarter of expenses.
- Energy buffer: at least one week of rest per cycle.
- Decision buffer: delay irreversible choices until patterns are clear.

Write these into your operating rhythm. Buffers make stability feel like breathing, not holding your breath.

4. Seed Bank for Creativity

Keep a simple "Idea Bank." Every undeveloped concept gets planted there.
Review it each quarter, old ideas often germinate when conditions change.

This transforms fleeting inspiration into a renewable resource.

Case Reflection: Ecosia, Storing Profit as Planetary Energy

Ecosia, the search engine that plants trees, captures digital attention and translates it into ecological yield.
Their business model **stores surplus energy as regeneration**: every search funds reforestation.

But their real innovation lies in governance.
They legally locked the company's ownership so profits can never be sold or extracted, permanently embedding "catch and store energy" into their structure.

That's the ultimate form of storage: protecting purpose from future depletion.

Takeaway

Growth without storage is just burnout on delay.
The goal isn't to collect everything, it's to **retain what sustains you.**

When your company learns to catch and store energy, financial, human, creative, you break the cycle of exhaustion and rediscover compounding vitality.

Stored energy gives you time to think, to rest, to evolve.

In forests, water storage means survival.
In business, stored energy means freedom.

Up Next

Chapter 4. Obtain a Yield, Design for Sustainable Profit
You've learned how to capture energy. Next, we'll explore how to turn that energy into yield, multiple, renewable forms of profit that feed your ecosystem instead of depleting it.

Chapter 4, Obtain a Yield: Design for Sustainable Profit

From Extraction to Regeneration in the Flow of Value

The Food Forest Lesson

Walk through a mature food forest and you'll see seven layers of life, all producing value at once, canopy, understory, shrubs, herbs, ground cover, roots and vines.
Each layer yields something different: fruit, shade, nitrogen, mulch, habitat.

No single part dominates.
The system's genius lies in **mutual productivity**, everything feeds something else.

That's the model for regenerative profit.
Every part of your company can yield, if you design for it.

In traditional business, profit means extraction: taking more value out than you put in.
In regenerative business, profit means **circulation**, the steady return of energy to sustain and expand life.

The question shifts from *"How much can we get?"* to *"How many kinds of value can we grow?"*

The Ecology of Yield

In permaculture, yield is defined not just as harvestable goods, but as *any form of energy stored for future use.*
A compost pile is yield. So is clean water, fertile soil, or biodiversity.

In business, the same pattern holds. Profit is not the only yield, it's one among many.
Regenerative companies design for **multiple simultaneous yields**, each feeding the next:

Yield Type	Description	Example
Financial Yield	Stable, fair profit-sustaining operations	Recurring revenue, long-term clients
Social Yield	Trust, goodwill and community reciprocity	Brand advocacy, partnerships
Human Yield	Learning, growth and fulfilment	Staff retention, creative capacity
Cultural Yield	Shared story, identity and purpose	Strong culture, belonging
Ecological Yield	Environmental repair or reduced harm	Circular production, offset systems
Intellectual Yield	Knowledge, IP, reusable frameworks	Courses, toolkits, publications

Each yield strengthens the system. Together, they create *true sustainability*, the ability to thrive without burnout or depletion.

The Myth of Single-Yield Business

The industrial business model rewards **narrow focus**. Specialize. Maximise one yield, usually financial and optimise everything around it.

That works — until the single yield collapses.

Consider monoculture farming: one crop, one goal, one disease away from failure.
The same fragility shows up in companies reliant on one product, one client, one founder or one market trend.

Regenerative founders diversify yields by asking:

> "What else can this create?"

A product that yields revenue *and* learning.
A meeting that yields decisions *and* trust.
A marketing campaign that yields customers *and* community.

When yield becomes multidimensional, your company becomes resilient.

Case Example: Tony's Chocolonely, Yield as Ethical Market Transformation

Tony's Chocolonely started as a protest, a journalist's attempt to expose child labor in chocolate.
Today, it's a thriving company proving that profit and justice can coexist.

Their design principle? **Yield as transformation.**
Every sale funds a transparent supply chain that rewards farmers directly.

They publish full supply chain data, reinvest in farmer equity and turn customers into activists.

The result:

- Financial yield through strong sales.
- Social yield through awareness and activism.
- Cultural yield through a bold, purpose-driven brand.
- Ecological yield through direct trade and reforestation.

Every layer reinforces the others. The company grows because its ecosystem thrives.

Designing for Multiple Yields

To design for regenerative yield, start with three shifts:

1. Define Success as a Cycle, Not a Point

Most companies treat profit as an endpoint, a finish line.
In living systems, yield is part of a cycle: growth → harvest → renewal.

Action:
At year-end reviews, don't ask, "Did we hit target?"
Ask, "What did this year's yield enable next year?"

Profit that doesn't feed renewal is sterile.

2. Stack Functions

In a food forest, one element serves many functions. A tree provides fruit, shade, mulch and microclimate regulation.

In business, the same principle applies. Each activity should produce **more than one yield**.

Example:

- A team retreat yields rest (human), new ideas (intellectual) and content (marketing).
- A partnership yields revenue (financial), visibility (social) and shared learning (intellectual).
- A hiring process yields talent (human), culture stories

(cultural) and operational clarity (structural).

If one activity doesn't yield at least two forms of value, redesign it.

3. Close the Loops

Regenerative yield requires circulation, nothing wasted.

If an activity creates value that isn't captured or reused, it leaks. Design your systems so that every output becomes input somewhere else.

Example:

- Customer feedback fuels product iteration.
- Internal learnings become marketing content.
- Offcuts from production become new products.

Interface Carpets turned industrial waste into material for new flooring. The result? A profitable circular loop that saved millions.

Every yield that loops strengthens the ecosystem.

Case Example: Dr. Bronner's, The Yield Multiplier

Dr. Bronner's produces soap, but its true product is **integrity**. They reinvest profits into organic agriculture, fair-trade partnerships and social justice campaigns.

Their yields are layered:

- **Financial:** profitable global brand.
- **Human:** staff ownership and equitable wages.
- **Social:** activism and trust.
- **Ecological:** regenerative farming supply chains.
- **Cultural:** a living example of ethical capitalism.

Each yield amplifies the next, customers buy not only soap, but story, justice and belonging.

Their financial health *depends on* their moral health, a regenerative loop at scale.

The Pattern of Regenerative Profit

Regenerative profit is the flow of life through a company, not accumulation, but recirculation.

Here's the pattern:

1. **Energy enters** through attention, money and human effort.
2. **Systems capture and store** that energy (Chapter 3).
3. **Designs obtain yield**, converting energy into multiple valuable outputs.
4. **Surplus circulates** through the ecosystem, feeding the next cycle (Chapter 1 ethics).

Profit becomes process, not prize.

The Founder's Role: From Harvester to Gardener

Traditional founders chase yield by harvesting more.
Regenerative founders design conditions where yield emerges naturally.

That shift changes everything:

- From extraction to invitation.
- From control to cultivation.
- From urgency to patience.

You no longer "drive" growth. You **tend** it, through soil health (culture), water flow (finance) and sunlight (purpose).

Profit becomes the forest floor: abundant, quiet, steady.

Case Example: Zingerman's, Yield Through Distributed Ownership

Zingerman's Community of Businesses (Ann Arbor, USA) demonstrates regenerative yield in human form.

Each new venture, bakery, creamery, roastery, is co-owned by its managing partners.
Profits stay local, circulating within the community rather than extracting upward.

Yield flows in multiple directions:

- Owners gain autonomy.
- Staff gain opportunity.
- Customers gain diversity and consistency.
- The town gains resilience.

Zingerman's teaches: the fairest share produces the longest yield.

From Burnout to Biodiversity

Founders often mistake burnout for weakness.
It's actually a **sign of yield imbalance**, too much extraction, too little renewal.

When your company yields only money, you deplete every other form of capital.
When it yields learning, rest and belonging as well, energy multiplies.

Diversity of yield equals diversity of resilience.

Field Guide: Designing for Yield

1. Map Your Current Yields

List all the value your company produces, tangible and intangible.
Use the six categories: Financial, Social, Human, Cultural, Ecological, Intellectual.
Mark which yields are strong, weak, or missing.

You'll likely see overemphasis on financial and intellectual yields, and neglect of human or ecological ones.
That imbalance points to your design opportunity.

2. Identify Your Keystone Yield

Every ecosystem has a keystone species, a source that supports the rest.
In your company, identify which yield sustains the others.

Example:

- In a creative agency, **learning** may drive innovation (financial) and morale (human).
- In a farm-to-table business, **soil health** may drive product quality (financial) and brand story (cultural).

Protect your keystone yield first.

3. Stack Functions in One Project

Take a major initiative and ask:

> "How could this generate at least three yields?"

Example: A product launch could yield:

- Revenue (financial)
- Thought leadership (intellectual)

- Partnership goodwill (social)
- Employee pride (human)

Write these yields into your plan, measure them, celebrate them, reinvest them.

4. Design Surplus Circulation

Decide how surplus flows.

- What percentage of profit feeds regeneration?
- What portion of time or creativity gets shared back?
- How will you make these loops visible to the team and community?

Transparency turns circulation into culture.

5. Create a Yield Dashboard

Build a simple dashboard tracking diverse yields.
Each quarter, record one key indicator per yield type, even qualitative data (stories, testimonials, wellbeing).

This reminds everyone: success is systemic, not singular.

Takeaway

Profit is a living relationship between energy and renewal.

When you design for yield, not just income, your company becomes fertile ground.
Every effort compounds, every surplus feeds the next cycle.

The forest doesn't need permission to prosper. It just knows how to yield.

So can you.

Up Next

Chapter 5, Apply Self-Regulation and Accept Feedback: Create Adaptive Systems.
We'll explore how regenerative companies use feedback, not fear, to stay alive, balanced and continually evolving.

Chapter 5, Apply Self-Regulation and Accept Feedback: Create Adaptive Systems

How Living Companies Stay Balanced Without Burnout

The Beehive Lesson

Walk near a beehive on a warm morning and you'll hear a steady hum, thousands of bees working in perfect rhythm.
No boss gives orders. No spreadsheet dictates flow.

When the temperature rises, some bees fan their wings to cool the hive. When it drops, others cluster tight to conserve heat. When nectar is scarce, foragers adjust their search patterns automatically.

The hive stays balanced because it's built on feedback, thousands of small, continuous corrections.

That's what self-regulation looks like in nature: no central control, just *responsive intelligence.*

In companies, this is what separates burnout-driven growth from regenerative stability.
Most organisations rely on **top-down control** or **delayed feedback**, quarterly reports, annual reviews, sudden crises. By the time information moves up the chain, it's too late to adapt.

Regenerative companies learn from the hive. They design systems where **feedback is frequent, local and kind.**

Why Control Fails

The industrial mindset treats control as the path to consistency: design rules, monitor performance, enforce compliance.
It works in machines but breaks in living systems.

Control creates fragility. It discourages experimentation, silences weak signals and punishes honest feedback.

The result:

- Problems hide until they explode.
- Innovation dies under bureaucracy.

- Leaders burn out trying to push order onto chaos.

Self-regulation doesn't mean absence of structure. It means structure that can flex, systems that *listen.*

Principle: Feedback Loops Create Stability

In ecology, every stable system depends on feedback loops: predator-prey cycles, nutrient exchange, climate regulation. Feedback doesn't prevent disturbance, it turns disturbance into learning.

A regenerative company designs similar loops at every level:

System Level	Example of Feedback	Adaptive Response
Individual	Regular check-ins and energy reflections	Adjust workload, clarify priorities
Team	Retrospectives, project reviews	Redesign process or rhythm
Customer	Real-time satisfaction metrics, story collection	Refine product or service
Financial	Monthly cash-flow reviews	Reallocate resources
Cultural	Anonymous pulse surveys, open Q&As	Evolve values and rituals

These loops transform feedback from anxiety into information. The goal isn't to prevent mistakes, it's to make them *smaller and faster.*

Case Example: Buffer, Radical Transparency as Self-Regulation

Buffer's internal operations are public: salaries, equity, revenue, even meeting notes.

This radical transparency replaces managerial policing with social feedback.
When everyone can see decisions, the system self-regulates, not through fear, but through visibility.

If salaries drift off balance, people ask why.
If priorities misalign, patterns surface quickly.

Transparency turns feedback from punishment into participation.

The takeaway:
The healthiest systems regulate themselves through visibility, not control.

The Ecology of Feedback

In natural systems, feedback has two directions: **negative** (corrective) and **positive** (amplifying).
Both are necessary.

- **Negative feedback** keeps balance, like sweating when you overheat.

- **Positive feedback** drives growth, like mycelium spreading where nutrients are rich.

In business:

- Negative feedback = early warnings (declining morale, customer churn).

- Positive feedback = signs of vitality (repeat clients, creative energy).

Regenerative founders design to notice both, and act before either becomes extreme.

Creating Adaptive Governance

Traditional governance assumes predictability.
Regenerative governance assumes change.

That means replacing rigid rules with **responsive frameworks.**

1. Distributed Decision-Making

Empower teams to decide within clear boundaries, like cells in a body responding locally.
Haier, the Chinese appliance company, turned 60,000 employees into 4,000 micro-entrepreneurs with full autonomy over small business units.

Result: faster response, higher innovation, no central choke-points.

2. Short Feedback Cycles

Instead of annual reviews, use short learning loops.
Basecamp runs six-week work cycles followed by a cooldown week for reflection. The rhythm itself is feedback, effort and rest, output and insight.

3. Transparent Metrics

Everyone should see the same data.
Transparency turns numbers into conversation, not control.

Case Example: Morning Star Company, The Self-Managing Tomato Plant

Morning Star, the world's largest tomato processor, runs without traditional managers.
Each employee has a **Colleague Letter of Understanding (CLOU)**, a written agreement describing their responsibilities and how others can hold them accountable.

Instead of hierarchy, they use **peer feedback and mutual commitment** as self-regulation.
If a worker sees waste, they have authority to fix it.
If performance slips, peers address it directly.

Productivity and morale both thrive because regulation happens *within* the system, not *above* it.

Morning Star proves: when feedback is shared, control becomes unnecessary.

Designing Feedback as Flow

Feedback shouldn't be an event; it should be a flow, built into daily operations.
The key is designing **lightweight, continuous signals** rather than heavy reviews.

Practical Feedback Channels

1. **Pulse surveys:** One-question weekly check-ins ("How sustainable does your workload feel?").

2. **Energy dashboards:** Teams track their energy like a metric, green (energised), yellow (tired), red (drained).

3. **Open retrospectives:** After each project, teams record insights publicly.

4. **Customer story wall:** Real feedback pinned in view, not hidden in reports.

5. **Financial dashboards:** Real-time cash visibility for all.

The goal is awareness, not surveillance.

Case Example: Zappos, Self-Regulation Through Culture

Zappos built a culture where customer service agents have full autonomy to delight customers, no scripts, no time limits.

Feedback flows directly from customer to employee.
When something goes wrong, agents adjust immediately rather than escalating up the chain.

Leadership focuses on designing the culture that supports this freedom: hiring for empathy, storytelling successes and celebrating recovery as much as success.

That's feedback ecology in action, learning built into every interaction.

Turning Feedback Into Learning

Observation (Chapter 2) gives you data.
Feedback turns that data into adaptation.

A regenerative founder creates rituals that transform feedback from emotion into insight.

1. Pause Before Reacting

When feedback arrives, don't defend. Ask, *"What pattern is this pointing to?"*
Separate the signal (truth) from the static (tone).

2. Close the Loop

If someone gives feedback, they should see the response.
This builds trust, people learn their voice matters.

3. Turn Feedback Into Design Inputs

Instead of fixing symptoms, redesign systems.
If people complain about meetings, maybe it's not the schedule, it's the purpose or decision-clarity.

Feedback reveals root causes if you listen at system level, not surface level.

Feedback and Founder Ego

Many founders struggle with feedback because it feels personal. But in living systems, feedback is neutral, it's just information.

Regenerative founders cultivate **ego composting**: letting criticism decompose into growth.

Practical habits:

- Invite feedback before it's forced.
- Celebrate the person who points out friction.
- Model humility, say "thank you" instead of "but."

The more you welcome feedback, the less crisis you'll face.

Field Guide: Building Self-Regulating Systems

1. Map Your Feedback Loops

Draw your company as a circle.
Identify the main flows of feedback: team → leadership, customer → product, finances → strategy.
Where are loops missing or delayed?
Add one lightweight mechanism for each missing loop.

2. Create Energy Metrics

Each month, ask your team to rate energy and focus on a simple 1-to-5 scale.
Graph it. Over time, you'll see seasonal patterns, when to push, when to rest.

Treat energy like cash flow, manage it actively.

3. Introduce Micro-Retrospectives

After any milestone, answer three questions:

- What worked?

- What drained energy?
- What will we adjust next time?
  Keep it under 15 minutes. The speed ensures the habit sticks.

4. Build a "Failure Compost"

Create a shared document where teams log small failures and what they learned.
Revisit quarterly. Patterns reveal systemic fixes.

In nature, decomposition feeds the next growth. In business, so does reflection.

5. Peer Accountability Agreements

Like Morning Star's CLOUs, write short agreements between colleagues outlining mutual expectations.
This shifts accountability from boss-employee to peer-peer, distributing regulation across the hive.

Case Reflection: Basecamp, Calm as a Feedback Signal

Basecamp uses "calm" as a metric.
If tension or urgency spikes, they treat it as a system signal, not a personal failure.

They adjust scope, clarify decisions, or pause releases.
By listening to stress as feedback, they prevent burnout before it starts.

Calm doesn't happen by luck; it's design.

Takeaway

Feedback is the pulse of a living company.
When you design for self-regulation, transparent data, local authority, continuous reflection, your business begins to behave like an ecosystem: adaptive, balanced, self-correcting.

Control is brittle.
Feedback is fluid.

The forest doesn't need a manager to stay in balance, it just needs to feel its own conditions.
So does your company.

Up Next

Chapter 6, Use and Value Renewable Resources: Regenerate, Don't Consume.
We'll explore how regenerative companies shift from depleting human and material energy to building renewal into every process, the business equivalent of fertile soil and flowing water.

Chapter 6, Use and Value Renewable Resources: Regenerate, Don't Consume

How to Build a Company That Gets Stronger the Longer It Runs

The Forest Floor Lesson

In a healthy forest, nothing is wasted.
Fallen leaves feed the soil. Decomposing wood becomes food for fungi. Every death becomes life for something else.

The system regenerates because it relies on **renewable resources**, sunlight, rainfall, microbial activity and time.
It doesn't depend on stored fossil energy or constant external inputs.

In contrast, most modern companies run like strip mines.
They extract energy, from people, materials, or attention, faster than it can be renewed.
The result is predictable: burnout, waste and brittleness disguised as success.

Regenerative businesses, like forests, **design for renewal.**
They favour systems that replenish themselves over those that require constant fueling.

What "Renewable" Really Means in Business

In ecology, a renewable resource is one that replenishes itself through natural cycles.
In business, that includes both **tangible and intangible forms of energy:**

Resource Type	Renewable Equivalent	Extractive Equivalent
Human Energy	Rest, creativity, trust, learning	Overwork, anxiety, burnout
Financial Energy	Steady, recurring income	Debt-fueled growth
Material Energy	Recycled or circular inputs	Disposable, linear production
Cultural Energy	Purpose, belonging, shared meaning	Fear, competition, secrecy
Social Energy	Collaboration, community goodwill	Exploitation, short-term gain

A regenerative founder's job is to replace extractive flows with renewable ones, one system at a time.

Efficiency vs. Renewal

Efficiency is doing more with less.
Renewal is ensuring there's always *more to give.*

The industrial economy rewards efficiency, squeezing every drop of time and output.
But efficiency without renewal is **a race to zero.**

Renewal asks different questions:

- Are people returning to work energised or empty?
- Are our products restoring ecosystems or degrading them?
- Is our culture generating trust faster than it spends it?

If efficiency extracts, renewal expands.

Case Example: Zingerman's, The Renewable Company

Zingerman's Community of Businesses has operated for over forty years without selling out or scaling unsustainably.
Their secret: **renewable systems of ownership and energy.**

- **Employee Ownership:** People share in profits and decisions, human energy renews itself through ownership.
- **Small and Slow Growth:** Each new business is

right-sized to its stewards, no overextension.

- **Cultural Compost:** Zingerman's constantly documents and teaches its methods, turning knowledge into reusable soil.

Their success isn't linear growth; it's **ongoing fertility.**

Designing for Human Energy Renewal

Human energy is the most renewable, and the most abused, resource in business.

You can't buy creativity or passion.
You can only cultivate the conditions where they regenerate naturally.

1. Rhythms of Rest

Every ecosystem has cycles, seasons, tides, day and night.
Businesses need them too.

Basecamp, for instance, runs six-week work cycles followed by two-week cooldowns.
The cooldown isn't "slack time"; it's composting time, reflection, cleanup and restoration.

2. Learning as Photosynthesis

Learning is how human systems convert experience into energy.
Encourage peer teaching, reflection circles and documentation.
Each lesson captured is sunlight stored for later.

3. Trust as Solar Power

Trust multiplies energy; suspicion drains it.
Renew trust through radical transparency, clear agreements and consistent follow-through.
It's the cheapest renewable resource there is, and the hardest to fake.

Case Example: Patagonia, Turning Material Flows into Renewal

Patagonia doesn't just minimise harm; it designs circular systems that create net-positive outcomes.

- **Repair Programs:** Customers can fix, not replace.
- **Recycled Inputs:** Over 70% of materials are post-consumer.
- **Ownership Transfer:** The company's profits now fund environmental repair forever.

Every loop closes back to renewal, of resources, of purpose, of legacy.
That's what business looks like when profit fuels regeneration.

The Hidden Renewable: Culture

Culture is a renewable resource, if designed that way.

A toxic culture burns energy; a regenerative culture generates it.
How? Through shared rituals, clear values and emotional honesty.

Example: At Dr. Bronner's, employee spiritual care, living wages and transparency create psychological safety.
People stay longer, innovate more and extend goodwill into every decision.

That's renewable morale, and it compounds faster than money.

From Ownership to Stewardship

Extractive businesses own things; regenerative businesses **care for them.**

Stewardship reframes success from possession to responsibility.
You don't "have" employees; you care for their wellbeing.
You don't "own" customers; you care for their trust.
You don't "control" nature; you care for the conditions that allow life to flourish.

Stewardship ensures every decision leaves more fertility than it found.

Designing Renewable Finance

Regenerative finance treats money as energy, not as a resource to hoard.

Principles of Renewable Finance

1. **Circulation over accumulation.** Keep money moving through fair wages, supplier support and reinvestment.

2. **Patient capital.** Seek investors aligned with time horizons of soil, not software.

3. **Debt as compost, not concrete.** Only borrow to build systems that renew themselves.

Example:
Ecosia legally bound itself to use profits for reforestation, permanently removing the temptation to extract.
Every euro spent becomes ecological yield.

That's money behaving like sunlight; constant, generous, renewable.

Designing for Regenerative Materials

Even product-based companies can design renewal into their physical systems:

- **Source renewably:** Choose materials that replenish naturally or are recycled.
- **Design for longevity:** Durable goods reduce extraction cycles.
- **Create take-back programs:** Turn customers into co-stewards.
- **Localise supply:** Proximity reduces waste and builds resilience.

Interface Carpets famously cut its environmental footprint by 96% by redesigning its entire supply chain around circular material flows, from fishing nets turned into carpet fibres to factories powered by renewables.

That's regeneration in motion.

Renewable Creativity

Creativity thrives under constraint and rest, not pressure. To keep it renewable:

- Protect **maker time**, uninterrupted blocks for deep work.
- Encourage **cross-pollination** between teams, diversity feeds ideas.
- Archive **creative compost**, reuse old ideas as inspiration.

When creativity is treated as infinite, it dies. When treated as a renewable crop, it multiplies.

Case Example: The "Regenerative Agency" Experiment

A small design studio in Copenhagen decided to test regenerative work cycles:
They shortened the workweek to four days, capped client load at three at a time and used Fridays for creative play and reflection.

After six months:

- Revenue held steady.
- Employee turnover dropped to zero.
- Client satisfaction rose.
- Burnout disappeared.

The team discovered the paradox of renewal:
By doing less, they created more value.

Their creativity became renewable.

Field Guide: Turning Consumption into Regeneration

1. Conduct a Resource Audit

List every major input your company relies on; material, human, digital, emotional.
For each, ask:

- Is it renewable or extractive?
- How could we close the loop or slow the flow?
- Who could we partner with to regenerate this input?

2. Design Rest Systems

Define rest as a function, not a perk.
Schedule seasons of reflection, silence, or light work.
Use these to integrate learning, compost failures and renew focus.

3. Track Human Energy Like a KPI

Create a simple "Energy Index."
Each team scores its weekly workload from 1–5 (drained to thriving).

Graph it over time, dips reveal structural issues faster than financial data.

4. Build Circular Flows

Wherever something leaves your system, money, material, or morale, ask how it might return.
Example: Offer product trade-ins, mentorship cycles, or customer reuse programs.

5. Regenerative Procurement

Choose partners and suppliers who share renewal ethics.
Buy from those who restore what they touch, local farms, ethical manufacturers, cooperative service providers.

Procurement becomes planetary design.

The Energy Equation

The simplest diagnostic question for any system:

Does this practice give more energy than it takes?

If not, it's extractive.
Regeneration begins when your net energy balance turns positive, in people, processes and ecosystems.

When renewal exceeds consumption, growth becomes effortless.

Case Reflection: Basecamp, Sustainable Calm

Basecamp's philosophy of "calm work" exemplifies renewable energy in culture.
They forbid long work hours, minimise meetings and focus on asynchronous communication.

The result?

- Fewer features but higher quality.
- Less stress but stronger loyalty.
- Steady profits with minimal marketing.

Their energy curve is flat, not spiky, and that's why they endure.

Calm is not the absence of ambition. It's the presence of renewal.

Takeaway

Nature doesn't hustle to survive. It regenerates through rhythm, rest and reciprocity.

When you design your company around renewable resources, human, financial, ecological, you break free from the tyranny of depletion.

You create a system that grows stronger the longer it runs.

The forest doesn't grow by consuming itself. Neither should your business.

Up Next

Chapter 7, Produce No Waste: The Circular Company. We'll explore how regenerative companies turn byproducts, time, data, materials, even mistakes, into resources that feed the next cycle of growth.

Chapter 7, Produce No Waste: The Circular Company

Turning Byproducts into Value in a Living Business

The Lesson of the Fallen Leaf

In autumn, a tree drops its leaves, millions of them.
To the untrained eye, it looks like loss. Decay. Waste.

But beneath the forest floor, those leaves begin the next season's fertility.
They feed fungi, shelter insects and return nutrients to soil.

In nature, **waste doesn't exist.**
Every output becomes input for something else.

In business, we've forgotten this.
We burn time, discard ideas, churn through people and call it "the cost of doing business."
But what if every form of waste could be composted, transformed into value?

That's the promise of the circular company.

Waste Is Just an Untapped Resource

Bill Mollison, co-founder of permaculture, wrote:

> "Waste is merely a resource in the wrong place."

In nature, there is no "away." Everything cycles.
In companies, waste is what happens when flows are **linear** instead of **circular**, when value moves in one direction and stops.

Common Forms of Business Waste

Type	Description	Regenerative Redesign
Material Waste	Overproduction, packaging, single-use inputs	Circular design, reuse, repair, recycling loops
Energy Waste	Overwork, misaligned effort, redundant processes	Process simplification, automation, rest cycles
Financial Waste	Unused assets, high churn, unmeasured returns	Resource audits, subscription or reuse models
Human Waste	Burnout, underused skills, lost trust	Job redesign, participatory governance, culture care
Creative Waste	Abandoned ideas, duplicated effort	Knowledge libraries, "idea compost" systems
Social Waste	Broken partnerships, ignored networks	Relationship loops, alumni and customer communities

When you learn to recognise these patterns, you stop patching symptoms and start redesigning systems.

Case Example: Interface Carpets, Industrial Ecology in Action

In the 1990s, Interface's CEO, Ray Anderson, realised the carpet industry was one of the most wasteful on Earth, mountains of petroleum-based nylon dumped into landfills each year.

He asked a simple question: *What if our factory worked like a forest?*

The company redesigned every process:

- **Modular carpets** allowed individual tile replacement instead of full replacement.
- **Recycled nylon** replaced virgin plastic.
- **Closed-loop factories** reused water and heat.

Over 25 years, Interface cut waste by 96%, tripled profit margins and became carbon-negative.

They didn't fix waste; they eliminated the concept of it.
Their byproducts became their business model.

The Organisational Compost Pile

Regenerative companies treat mistakes, friction and redundancy as compost, not trash.

> Compost isn't glamorous, but it's fertile.
> It's where transformation happens quietly.

Every problem contains nutrients for the next innovation, if you're willing to decompose it.

Composting in Practice

- When a project fails, **document the learning.**

- When a client leaves, **collect their honest feedback.**
- When an employee exits, **harvest their insights** in a closing reflection.

Over time, your organisation develops **fertile memory**, rich soil for better design.

Case Example: Zappos, Customer Waste as Opportunity

Zappos built its reputation not on perfection but on recovery. Every return or complaint is treated as a chance to deepen trust, not a loss.

They redesigned "returns" into "relationship moments." Instead of outsourcing customer service, they made it a core competence, agents empowered to delight, not defend.

The waste of dissatisfaction became a renewable resource: loyalty.

Zappos demonstrates that regeneration often starts where others see expense.

The Hidden Waste of Human Energy

The most overlooked form of waste in business is **emotional waste**, burnout, disengagement, or cynicism.
You can measure it by the number of people doing work that no longer nourishes them.

A regenerative company prevents this by **closing human energy loops.**

1. Feedback as Fertiliser

Use short reflection cycles to catch frustration early and compost it into improvement.

2. Role Composting

When roles stagnate, compost them. Retire old titles, evolve responsibilities and replant people where their energy grows again.

3. Cultural Renewal

Every few years, revisit company values through stories, not slogans. Let them evolve like an ecosystem, not a monument.

Waste occurs when you treat people as static; renewal happens when you treat them as living.

Creative Waste and the Idea Bank

In creative companies, ideas often die unseen, deleted drafts, abandoned campaigns, half-built products.
That's creative topsoil washing away.

A regenerative approach builds **Idea Banks**, shared compost heaps for unfinished concepts.

- Every draft or experiment is logged with a short note: "Why we paused."

- Every quarter, revisit the compost pile. Often, old ideas sprout under new conditions.

This is how innovation becomes cyclical, not forced, but emergent.

Case Example: Basecamp, Efficiency Through Simplicity

Basecamp combats waste through minimalism.
They avoid feature creep, meetings and overcommitment.
Every process is questioned: "Does this give more energy than it takes?"

Their radical simplicity eliminates time waste, cognitive overload and morale decay.

Waste reduction at Basecamp isn't about doing more, it's about doing *only what matters.*
Their simplicity is their sustainability.

Circular Design Thinking

To eliminate waste, redesign your systems around cycles, not lines.

The Linear Model

Input → Use → Output → Disposal

The Circular Model

Input → Use → Output → **Reintegration** → Renewal

Ask: *Where does this end up?*
If the answer is "somewhere else," you have waste.
If the answer is "back into our ecosystem," you have regeneration.

Case Example: Toast Ale, Brewing Beer from Bread Waste

Every year, millions of loaves of bread are discarded in the UK. Toast Ale uses that bread to brew beer.

Their slogan: "Raise a Toast to Change."
By closing a single waste loop, they created:

- A profitable product
- A movement around food waste awareness
- Funding for food charities

Toast Ale shows that waste isn't just avoidable, it's a source of innovation and identity.

The Waste Audit

Observation and feedback (from earlier chapters) help you see where energy leaks.
A **waste audit** makes it visible, the first step in circular design.

Conduct one quarterly:

1. List all resources (time, materials, money, human energy).
2. For each, track where it exits the system.
3. Ask: *Can this output feed another process?*

Example:

- Content marketing creates unused assets → convert to training resources.
- Product offcuts → turn into samples or upcycled goods.
- Employee downtime → invest in skill-building.

Waste reveals design opportunity.

From Guilt to Curiosity

Many sustainability efforts fail because they're driven by guilt, "We should waste less."
Regenerative design is driven by curiosity: *"What else could this become?"*

Guilt paralyses; curiosity creates.
The mindset shift from reduction to regeneration changes the culture from fear to invention.

When you start seeing waste as potential, creativity explodes.

Field Guide: Turning Waste Into Value

1. Conduct a "System Compost" Workshop

Gather your team and list every form of waste, material, time, or emotional.
Ask three questions:

- What's the root cause?
- What nutrients does it contain? (learning, relationships, content, ideas)
- How could we compost it into something new?

Pick one to pilot this quarter. Document the results.

2. Build a Circular Asset Map

Map all your outputs: reports, content, prototypes, byproducts.
Draw arrows showing how they could feed another process.
Example: Webinar → blog series → onboarding material → public whitepaper.
Your business becomes an ecosystem of reuse.

3. Introduce a "Waste Ledger"

Track waste like expenses.
Each department logs its primary waste form and one monthly action to reduce or repurpose it.
Celebrate repurposing wins the same way you celebrate sales wins.

4. Create a "Failure Compost" Repository

Store postmortems of failed projects in one place.
Tag them by theme (timing, team, tech).
Review quarterly, you'll spot patterns faster and make better next moves.

5. Apply the 3-R Cycle: Reduce, Reuse, Regenerate

- **Reduce:** Simplify before optimising.
- **Reuse:** Capture before discarding.
- **Regenerate:** Design the next version to feed others.

Every process should include at least one regenerative step.

Case Reflection: The Circular Café

A café in Melbourne redesigned around zero waste.
Coffee grounds feed nearby community gardens.
Milk jugs are returned to suppliers.
Pastry scraps go into next-day desserts.
Customers get discounts for reusable cups.

Waste dropped by 90%.
But something else happened: customers felt part of the system.
They became co-regenerators.

The café discovered the true reward of producing no waste, *shared ownership.*

The Regenerative Equation

Waste = Disconnection.
Value = Connection.

Every time a system connects its outputs to another input, it gains vitality.
The tighter the loops, the longer the life.

The circular company doesn't eliminate waste by working harder.
It eliminates it by working wiser, designing every process as a living cycle.

Takeaway

Waste is a design invitation.
When you start seeing waste as potential, your company becomes a creative ecosystem, where every ending feeds a beginning.

The fallen leaf doesn't mourn. It becomes the soil for spring.
So can every part of your business.

Up Next

Chapter 8, Design from Patterns to Details: The Blueprint of Living Companies.
We'll explore how regenerative founders use *patterns*, recurring natural designs, to guide decisions before diving into details, ensuring coherence and simplicity as their organisations evolve.

Chapter 8, Design from Patterns to Details: The Blueprint of Living Companies

How Regenerative Founders Build Wholeness Before Efficiency

The River and the Pebble

Stand beside a river. Watch how the water flows, not in straight lines, but in patterns.

The bends, ripples and eddies aren't random. They form through consistent laws of movement, gravity and resistance.

Now, pick up a pebble. It's shaped by that same river, every swirl, every flood, every slow season. The river didn't design the pebble directly. It designed the pattern that shaped it.

This is how nature creates complexity without chaos.
It works from **patterns to details**, universal shapes (branching, spirals, waves) that adapt themselves automatically to local conditions.

In business, the same principle builds coherence and calm.

Most companies do the opposite: they dive into details first, job titles, tech stacks, procedures, without understanding the larger patterns. The result?
Complexity without connection.

Regenerative founders start by seeing the pattern, the flow of energy, value and relationship, before designing the specific form.

The pattern holds the logic of life.

Pattern Literacy: The Hidden Skill of Regenerative Design

Every living system has recognisable patterns:

- **Branching**: flow and distribution (trees, rivers, supply chains)
- **Webs**: connection and resilience (mycelium, networks, culture)

- **Spirals**: iterative growth and renewal (shells, learning loops, innovation)
- **Cells**: autonomy within structure (honeycomb, teams, franchises)

Understanding these gives founders a language for structure. Instead of reinventing strategy, you **borrow nature's blueprints.**

Examples of Business Patterns

Natural Pattern	Business Translation	Example
Branching	Flow of information, decisions and value	Decentralised leadership networks
Web	Collaboration and interdependence	Cross-functional teams
Spiral	Iterative learning and improvement	Agile cycles, design sprints
Cellular	Semi-autonomous units with shared DNA	Zingerman's community model

Patterns act like DNA, simple instructions that generate infinite diversity.
When you design your company from patterns, the details emerge naturally.

Case Example: Haier, The Cellular Organisation

Haier, one of the world's largest appliance manufacturers, transformed itself from a hierarchical bureaucracy into 4,000 self-managed micro-enterprises.

Each team of 10–20 people operates like a **cell**: autonomous, customer-facing and responsible for its own profit and loss. Shared values and strategy form the "DNA," but decisions happen locally.

This pattern mirrors biological life: cells self-regulate and specialise while remaining part of a larger organism.

Haier's growth is fractal, scalable without control, adaptive without chaos.

That's design from pattern to detail at global scale.

Pattern 1: The Branching Flow

A river system teaches efficient distribution: wide trunk → narrowing channels → fine roots.
It balances central coordination with local responsiveness.

In companies, **branching patterns** ensure flow without bottlenecks.

Applications:

- Communication systems that start broad, then narrow (all-hands → teams → pairs).
- Decision flows that cascade context, not commands.
- Financial systems where profits branch outward, reinvested in edges closest to customers.

When communication jams, check your branching.
If too top-heavy, flow stagnates; if too fragmented, it floods.

Healthy rivers, like healthy companies, keep movement continuous and proportional.

Pattern 2: The Web

A spider's web is light but strong, tension evenly distributed.
It doesn't resist disturbance; it absorbs and rebalances it.

Web-based companies emphasise relationships over hierarchy.
They strengthen connection density, not job titles.

Practical translation:

- Encourage cross-team mentorship and shadowing.
- Track network health, not just reporting lines.
- Reward collaboration loops where value circulates.

When people feel connected laterally, self-regulation improves naturally.
The web replaces bureaucracy with belonging.

Pattern 3: The Spiral

Spirals capture the rhythm of learning, progress through repetition, not linear ascent.
In nature, spirals appear in galaxies, shells, hurricanes; each iteration builds on the last.

In regenerative companies, the spiral is the pattern of *iteration.*

Example:

- Project debriefs evolve into templates.
- Customer feedback informs the next version.
- Knowledge compounds through cycles of reflection and adaptation.

This pattern keeps your company in perpetual learning mode, growing smarter without needing to grow faster.

Patagonia's product design operates like a spiral: refine, repair, redesign, every cycle increases quality and reduces impact.

Pattern 4: The Cell

Cells are the building blocks of all life, small, semi-autonomous units connected by shared DNA.

In business, this pattern appears as **fractal structure**, teams or ventures that mirror the company's values, but adapt locally.

Zingerman's uses this pattern perfectly.
Each new business, the bakehouse, creamery or roastery, shares the same purpose ("great food, great people, great service") but operates independently.

The result? Scale through replication, not control.
Each cell learns, adapts and regenerates energy, contributing to the larger organism.

Case Example: Dr. Bronner's, The Spiral of Purpose

Dr. Bronner's product line looks simple; soap.
But the company's impact spirals outward:

1. Clean products.
2. Ethical supply chains.
3. Farmer cooperatives.
4. Regenerative agriculture education.

Each layer feeds the next, compounding yield and influence. Their spiral of purpose keeps the business alive with renewal.

Dr. Bronner's doesn't chase new markets; it deepens existing ones, a pattern of circular expansion.

Pattern-Based Design Process

Designing from patterns doesn't mean ignoring details.
It means **sequencing design correctly**: see structure before tinkering.

Here's how regenerative founders work:

1. **Observe Patterns.** Identify how energy already flows, communication, money, morale.
2. **Name the Pattern.** Is this a river (flow), web (connection), spiral (learning), or cell (structure)?

3. **Design Within the Pattern.** Adjust scale, direction, or rhythm, not the essence.

4. **Add Detail Last.** Choose tools, policies, or metrics that support the pattern, not override it.

Patterns act like gravity: you don't control them, you cooperate with them.

Case Example: Buurtzorg, The Fractal Care Organisation

Buurtzorg, a Dutch healthcare company, operates through 900 small, self-managed nurse teams.
No middle management. Minimal central office.

Each team runs its own budget, schedules and client care, supported by a digital platform that shares best practices.
The pattern: **cellular autonomy + shared DNA (patient dignity).**

Patient satisfaction is the highest in the industry; costs are the lowest.

By designing from pattern (human-scale, trust-based care), Buurtzorg achieved what bureaucracy never could.

The Danger of Detail First

When companies design from details, tools, policies, technology — they create **accidental ecosystems** full of friction.
Every decision solves a local issue but undermines global coherence.

Symptoms:

- Fragmented software systems.
- Conflicting incentives.
- Repetitive meetings.
- Exhausted founders.

Details without pattern clarity are weeds, they multiply, not mature.

Patterns prune complexity by giving every detail a home.

Pattern Mapping Exercise

Think of your company as a landscape.
Ask:

- Where does flow need branching?
- Where does resilience need webbing?
- Where does learning need spiralling?
- Where does autonomy need cells?

Sketch your business in these shapes.
You'll instantly see where structure and purpose are misaligned.

This is design literacy, learning to see what wants to emerge.

Pattern Recognition in Decision-Making

Patterns make decisions faster.
When faced with choices, ask: *Which pattern does this reinforce?*

Example:

- Should we hire more managers? → Does that strengthen the web or choke the flow?
- Should we add another product? → Does it extend the spiral of learning or create fragmentation?
- Should we centralise operations? → Does that feed or starve the cells?

Decisions become simpler when you see them as pattern alignment, not preference.

Field Guide: Design from Patterns to Details

1. Identify Your Core Patterns

Gather your team and map your dominant organisational patterns:

- Communication flow (branching or web?)
- Learning rhythm (linear or spiral?)
- Structure (hierarchy or cellular?)
- Purpose (finite or evolving?)

Name the top three that define your company.

2. Create a Pattern Charter

Document these patterns as design principles.
Example:

> "We grow through spirals of learning, not linear scaling."
> "We organise as cells with autonomy within shared purpose."

This becomes your blueprint for decision-making.

3. Align Details with Patterns

Audit your current systems.
For each process, ask:

- Which pattern is this serving?
- Which pattern is it violating?
  Then adjust accordingly.

4. Teach Pattern Language

Train your team to use natural metaphors, flow, edge, root, canopy.
It builds intuitive alignment.
Once everyone speaks this shared language, coordination happens naturally.

5. Design the Next Layer

After patterns are clear, add details: roles, tools, metrics, visuals.
Details should make patterns visible, not invisible.
If a system feels heavy, you've lost the pattern.

Case Reflection: Basecamp, The Pattern of Calm Simplicity

Basecamp's entire model follows a simple pattern: *constraint breeds creativity.*
Small teams, limited scope, calm communication.
Every detail (tool, policy, workflow) aligns with this pattern.

That's why they've stayed small, profitable and focused for decades.
It's not restraint; it's rhythm.

Takeaway

Patterns are nature's shortcuts, frameworks that turn chaos into coherence.

When you design your company from patterns to details, you create elegance that lasts.
Every policy becomes a leaf, every team a branch, every value a root, all connected by invisible order.

The forest doesn't micromanage its leaves.
It follows patterns that work.

So can you.

Up Next

Chapter 9, Integrate Rather Than Segregate: The Power of Connection.

We'll explore how regenerative companies dissolve silos, strengthen interdependence and design cultures where collaboration replaces competition, just as ecosystems thrive through integration, not isolation.

Chapter 9, Integrate Rather Than Segregate: The Power of Connection

How to Build Companies That Thrive on Relationship, Not Rivalry

The Lesson of the Forest Edge

Walk along the edge of a forest where it meets a meadow. Notice how the air hums with life. You'll see butterflies, bees, mushrooms and wildflowers that don't exist deep inside the woods or in the open field.

Ecologists call this the **edge effect**, the fertility created where two ecosystems meet.
At these intersections, diversity overlaps, nutrients exchange and new possibilities emerge.

Nature integrates.
The most life happens in the spaces between.

Companies, unfortunately, are built the opposite way.
We separate departments, functions and people into neat boxes, marketing, finance, operations, HR. Each optimised for efficiency. Each isolated from others.

The result: dead zones between the silos.
Information stops flowing. Energy dissipates. Innovation dies quietly.

Regenerative companies reverse this pattern.
They design for edges, not walls.
Integration becomes their advantage.

Segregation: The Industrial Legacy

Industrial management was built on control and separation.
The logic was simple: divide tasks, measure output, reward individually.

It worked for machines, not for ecosystems.

Segregation creates predictable side effects:

- Duplicated effort and wasted energy

- Hidden knowledge and slow decisions
- Burnout through isolation
- Innovation limited to one department

It's the equivalent of farming a monoculture, easier to manage, but more fragile.

Integration, by contrast, is the logic of forests and mycelium, decentralised, diverse and deeply communicative.

The Ecology of Integration

In natural systems, integration happens through **mutual benefit** and **feedback loops.**

- Trees exchange nutrients through fungal networks.
- Birds spread seeds for fruiting plants.
- Predator-prey relationships maintain population balance.

Integration is not harmony; it's dynamic tension, difference working in rhythm.

In business, integration means creating **structures and cultures that reward collaboration and feedback across boundaries.**

It's the difference between a mechanical company (efficient but brittle) and a living company (adaptive and resilient).

Case Example: Zingerman's, Community of Businesses

Zingerman's is not one company but a **network of interdependent ventures**: bakery, creamery, coffee roastery, restaurant group, training arm.

Each has its own leadership and profit centre, but they share vision, values and governance.

Integration mechanisms:

- Shared ownership and brand.
- Cross-company profit sharing.
- Rotating leadership councils.
- Open-book finance across ventures.

Result: collective stability, individual creativity.
Each business supports the others like species in a forest.

Their motto says it all: "*We share energy, not control.*"

Integration Starts with Perspective

To integrate, you must see your company as a **whole system**, not a collection of parts.

Think of these layers as interdependent:

1. **Purpose (Root)**, shared reason for being.

2. **People (Trunk)**, relationships and culture.

3. **Processes (Branches)**, how work and information flow.

4. **Products (Leaves)**, expressions of the whole system.

When alignment breaks at any level, waste appears, just like a tree with blocked sap flow.

Integration means tending to all layers simultaneously, ensuring flow and feedback connect root to leaf.

Case Example: Patagonia, Integration of Ethics and Operations

At Patagonia, ethics aren't marketing, they're infrastructure. The company integrates sustainability into product design, supply chains, HR and finance.

- Designers collaborate with environmental scientists.
- Employees can take paid activism leave.
- Financial systems prioritise long-term stewardship.

This full-stack integration ensures consistency: purpose flows through every decision.

The lesson:
Integration is not alignment meetings, it's *shared DNA.*

The Edge Principle

Innovation lives at edges, the meeting points between differences.
When you intentionally connect diverse perspectives, disciplines, or roles, you generate adaptive intelligence.

Design Edges in Your Company

- Cross-functional teams (finance + design + operations).
- Rotating roles that expose people to different ecosystems.
- Regular "edge conversations" between departments that rarely meet.
- Co-design sessions with customers and suppliers.

The goal isn't consensus; it's *cross-pollination.*

Every edge is a potential ecosystem.
The more you integrate edges, the more alive your organisation becomes.

Case Example: IDEO, The Edge as Innovation Engine

IDEO, the design firm, built its reputation by integrating disciplines, engineers, psychologists, anthropologists and designers working side by side.

Their process thrives on tension.
When a product engineer sits beside a behavioural scientist, creativity multiplies.

They call it "radical collaboration."
Each discipline sees blind spots that others miss.

That's not diversity for optics, that's **ecological integration**.

Integration in Communication

Communication systems are the nervous system of a company. Most are designed linearly, top-down memos or isolated chats. Regenerative systems are **networked**, information flows both ways, continuously.

Healthy Communication Design

1. **Transparency:** Shared dashboards, open books, visible progress.

2. **Proximity:** Conversations close to where information originates.

3. **Rhythm:** Regular touchpoints without overload.

4. **Psychological Safety:** People can share truth without fear.

When communication integrates, culture becomes self-correcting, less gossip, more growth.

Integration Through Governance

Centralised control fragments energy.
Integrated governance distributes power while maintaining coherence.

Patterns That Enable Integration

- **Sociocracy:** Circles of roles linked by feedback loops.
- **Holacracy:** Distributed authority through clear domains.
- **Advisory Councils:** Rotating teams bridging strategy and operations.
- **Participatory Budgeting:** Everyone contributes to resource allocation.

Each model mirrors ecosystems; multiple centres of intelligence, continuous feedback, shared purpose.

Case Example: Buurtzorg, Integrated Autonomy

Buurtzorg's nurses operate in self-managing teams, but they share digital systems, training and culture.
They integrate autonomy (local decisions) with alignment (shared purpose).

When integration is strong, autonomy amplifies it.
When it's weak, autonomy fragments it.

The lesson: **integration enables decentralisation.**

Integration Beyond the Company

Regenerative companies extend integration outward, blending business, community and ecology.

Ecosia integrates digital profit and planetary renewal: every search funds tree planting.
Allbirds integrates design, fashion and carbon accounting into one seamless story.
Local B Corps collaborate to build shared supplier ecosystems.

Integration scales regeneration beyond walls, into communities and supply chains.

Field Guide: Designing Integration

1. Map Your Edges

Draw your organisation as overlapping circles: departments, products, partners.
Label where they interact, and where they don't.
Edges without exchange are opportunities for innovation.

2. Build Cross-Pollination Rituals

- Host "Edge Dinners", informal gatherings mixing disciplines.
- Pair unexpected collaborators for one-day sprints.
- Rotate observers across departments.

Small connections build massive resilience.

3. Integrate Purpose Into Every Decision

Ask: *How does this decision express our core purpose?*
If the connection isn't clear, pause.
Every misaligned detail drains energy from the system.

4. Co-Own Problems

When challenges arise, resist assigning them to one department. Instead, form temporary "integration pods" that cut across functions to diagnose and solve together.

This keeps ownership shared, not siloed.

5. Feedback Loops Between Stakeholders

Design mechanisms that connect your ecosystem:

- Customer feedback → product iteration → marketing storytelling → supply redesign.
- Supplier feedback → quality improvement → employee education. Every loop shortens reaction time and increases intelligence.

Case Reflection: Interface Carpets, Full-System Integration

Interface integrated sustainability across every level:

- Supply chain reimagined for zero waste.
- Employee culture trained in ecological literacy.
- Profit reinvested in renewable design.

Their success wasn't just environmental; it was relational.
Every person saw their role in the larger pattern.

That's the goal of integration: *everyone knows how their work feeds the whole.*

Integration as Leadership Practice

A regenerative founder is less a commander and more a **pollinator**.
They move between edges, connecting insights and people.

Their questions shape culture:

- Who isn't in this conversation?
- What connections are we missing?
- How can this system feed itself?

Integration begins not with new tools but with new attention.

Takeaway

Segregation creates efficiency; integration creates life.
A regenerative company is an ecosystem, interconnected, responsive and alive with edges of collaboration.

Your role as founder is not to manage every detail, but to design conditions where connection happens naturally.

The forest doesn't optimise its trees; it integrates them.
That's why it endures.

Up Next

Chapter 10, Use Small and Slow Solutions: The Pace of Living Growth.
We'll explore how regenerative founders resist the pressure to scale fast and instead build patiently, using small, steady steps that compound over time like soil fertility or coral reefs.

Chapter 10, Use Small and Slow Solutions: The Pace of Living Growth

How to Build a Company That Compounds Like Soil, Not Like a Stock Chart

The Lesson of the Soil

If you dig a handful of soil from a healthy forest, you're holding billions of living organisms, fungi, bacteria, nematodes, roots. None of them move fast.
Yet together, they create fertility that can last for centuries.

That's how nature scales: slowly, iteratively, through compounding relationships.

Speed in nature is often a survival mechanism; longevity is a design choice.
Forests, reefs and prairies evolve through **small, slow, cumulative solutions** that create resilience.

Human enterprises could do the same, if we stopped mistaking motion for progress.

The Cultural Myth of "Faster is Better"

Modern entrepreneurship inherited an industrial tempo: fast growth, fast funding, fast exits.
We're taught to scale before we understand, automate before we stabilise and sprint before we can walk.

But nature shows that speed without structure leads to collapse. Trees that grow too fast fall in storms. Rivers that rush too violently erode their banks.

Most founders discover this too late, after energy, culture, or clarity have washed away.

Regenerative founders design for **steady compounding growth** instead of exponential extraction.

Why Slow Is Smart

"Slow" in regenerative design doesn't mean lazy or cautious.
It means *appropriately paced*, growth that matches available energy and context.

Ecological Translation

- **Fast systems** extract energy.
- **Slow systems** recycle it.

In business terms:

- Fast growth consumes attention, capital and people.
- Slow growth allows feedback, integration and trust to build in sync.

A slow company learns faster because it has time to notice.

Case Example: Basecamp, The Calm Company

Basecamp (formerly 37signals) has deliberately stayed small for over two decades.
They reject the startup growth narrative, focusing instead on **sustainable profitability and calm work.**

Their small-and-slow design includes:

- Limited product scope.
- Six-week cycles for deep focus.
- Intentional constraint as innovation fuel.

They don't scale headcount or speed, they scale *clarity.*

The result: decades of consistent profit, loyal employees and steady creativity.
Patience becomes their competitive edge.

The Pattern of Living Growth

In nature, healthy growth follows a simple sequence:

1. **Establish roots.** Observation, connection, stability.
2. **Send shoots.** Experiments, prototypes, relationships.
3. **Spread canopy.** Expansion through proven patterns.
4. **Drop seeds.** Replication through mentorship, open knowledge, or spin-offs.

Each stage builds fertility for the next.
Skip one and collapse follows.

Regenerative founders grow this way, deliberately, locally, cyclically.

Small Systems, Big Leverage

Small and slow solutions work because they leverage **local intelligence** and **continuous feedback**.

- A small team can adapt faster than a department.
- A small product can evolve faster than a suite.
- A small decision loop learns faster than a bureaucracy.

When you stay small long enough, you gather enough feedback to scale wisely, not blindly.

Nature's secret:

The smaller the experiment, the faster the learning.
The slower the rollout, the longer the life.

Case Example: Zingerman's, Fractal Expansion

Instead of franchising, Zingerman's chose to **replicate smallness.**
Every new business is an independent, co-owned "cell" with its own P&L and personality.

They call it *"scaling by staying small."*

Each business grows slowly enough to maintain culture, train leaders and test systems before expanding further.

Over forty years, this slow design produced a network of thriving ventures, none rushed, none burned out.

From Startup Speed to Seasonal Rhythm

Nature moves in seasons: growth, harvest, rest, renewal.
Businesses that ignore this rhythm burn out their ecosystems.

Regenerative founders build **seasonal pacing** into strategy:

- Growth seasons (launches, campaigns).
- Maintenance seasons (system building, documentation).
- Reflection seasons (pause, learning, composting).

Small and slow companies live in rhythm with their own energy. They don't force perpetual summer.

Case Example: Dr. Bronner's, The Long Game of Integrity

Dr. Bronner's could have sold out decades ago. Instead, they chose a 70-year slow build rooted in ethical supply chains.

Each relationship with a farming cooperative took years to establish.
But once in place, those networks became nearly unbreakable.

Patience became profit.
Integrity became infrastructure.

Their slow scaling created compounding yield, social, ecological and financial.

The Micro-Solution Strategy

Big problems can only be solved through small experiments repeated over time.

Regenerative founders apply this through a **micro-solution loop**:

1. Start with one pain point.

2. Design a minimal, reversible fix.

3. Observe results.

4. Adjust, then scale what works.

This keeps adaptation constant and stakes low; learning replaces risk.

Example:
Instead of overhauling an entire hiring system, start with one role and one new process.
Observe what changes, communication, energy, culture, then replicate.

That's evolution, not revolution.

The Economics of Slowness

Fast companies chase capital.
Slow companies accumulate *resilience capital*: trust, reputation, mastery and community goodwill.

These are slower to build but harder to destroy.

Patagonia's "slow capital" approach, funding its own growth through retained earnings, gave it total strategic freedom.
No investor dictates its pace.
The company grows at the speed of integrity.

Small and Slow as Design Constraint

Constraints focus creativity.
When you intentionally limit scope or pace, your team is forced to design smarter, lighter and more human systems.

Practical examples:

- Cap meetings per week.

- Limit open projects per person.
- Launch smaller versions before full releases.
- Choose the simplest tool that works.

Constraint becomes clarity.
You end up doing less, better, which compounds value over time.

Field Guide: Practising Small and Slow Solutions

1. Adopt Seasonal Planning

Plan your company year like a garden:

- **Spring:** Experiment and plant (new ideas).
- **Summer:** Focus and tend (execute).
- **Autumn:** Harvest and share (review, celebrate).
- **Winter:** Rest and compost (reflect, rebuild).

Slow down to the speed of seasons.

2. Shrink Your Experiment Size

Before any major initiative, ask:

> What's the smallest possible test of this idea?

Run it. Measure impact.
If it works, expand slightly. If it fails, compost the learning.

3. Create a "Slow Board"

Track initiatives over time instead of targets per quarter.
Visualise how small actions compound.
Encourage depth over speed, let patterns mature before layering new ones.

4. Limit Inputs

Just as ecosystems cap resource intake, businesses need energetic caps.
Set boundaries for projects, meetings, or hires.
Leave room for composting, the unseen work of integration and rest.

5. Celebrate Patience

Reward the work of consistency, not just novelty.
Highlight teams that improved a system over six months, not just those who launched something new.
This reframes culture from "launch mode" to "learning mode."

Case Reflection: Basecamp and the Power of Enough

At Basecamp, founders Jason Fried and David Heinemeier Hansson openly discuss their philosophy of *enoughness.*

They choose not to double growth if it means doubling stress.
They measure progress in quality of life and clarity of purpose.

Their slow business thrives because it's designed to last.
As Fried says:

> "You don't need to be bigger to be better. You need to be calmer to think clearly."

That's slow growth wisdom in one sentence.

Takeaway

Speed is seductive, but slowness is sovereign.
Small and slow solutions don't delay success; they deepen it.

Like soil, they build capacity for the long term: invisible, patient, alive.
When you align your company's pace with life's natural rhythm, growth becomes inevitable, and sustainable.

The forest doesn't rush.
It compounds quietly.

Up Next

Chapter 11, Use and Value Diversity: The Ecology of Creativity.
We'll explore how regenerative founders design diversity, of people, ideas, revenue and perspectives, as the ultimate resilience

strategy, drawing lessons from ecosystems where difference drives stability.

Chapter 11, Use and Value Diversity: The Ecology of Creativity

How Difference Becomes the Source of Resilience and Innovation

The Lesson of the Rainforest

A rainforest isn't stable because it's uniform, it's stable because it's complex.
Thousands of species coexist, each performing a different role. The orchid clings to branches, the ants defend trees, the fungi digest debris, the canopy filters sunlight.

No single organism dominates, yet the system thrives.
Every disturbance, a fallen tree, a flood, a dry spell, sparks regeneration through diversity.

That's the pattern of resilience: when one element fails, others adapt.

The same pattern applies to companies.
Uniform teams, narrow revenue streams and monocultural thinking create fragility.
Diversity, of people, products, ideas and models, makes a company anti-fragile.

The Monoculture Trap

Most organisations start diverse, different skills, backgrounds and energies united by passion.
But as they grow, efficiency takes over.

Processes standardise. Hiring narrows. Culture calcifies.
Before long, diversity becomes risk instead of resource.

That's the moment fragility sets in.

Like monocrop farming, it works fine until one pest, market change, tech shift, founder burnout, wipes it out.

Regenerative founders design against monoculture by embedding **diversity as structure**, not decoration.

The Four Dimensions of Business Diversity

1. Human Diversity, People and Perspectives

The foundation of all resilience.
Different lived experiences generate different solutions.

A team full of people who think alike can execute quickly but innovate poorly.
Diversity slows decisions in the short term and strengthens them in the long term.

2. Creative Diversity, Ideas and Expression

Multiple modes of thinking, analytical, intuitive, visual, relational, generate creative abundance.
When creativity feels stuck, it's usually because a single thinking style dominates.

3. Economic Diversity, Revenue and Resource Streams

Just as ecosystems rely on many nutrient flows, resilient companies rely on multiple sources of yield.
Diverse income streams stabilise operations through changing conditions.

4. Ecological Diversity, Supply, Partners and Impact

Integrating multiple suppliers, materials and ecosystems reduces dependency and increases adaptability.

Together, these layers create the regenerative equivalent of biodiversity, a living safety net woven from difference.

Case Example: Patagonia, Diversity as System Design

Patagonia's success isn't just about products.
It's about the diversity built into its entire ecosystem:

- **Product diversity:** technical gear, casual wear, repair services.
- **Cultural diversity:** activists, designers, scientists, craftspeople.
- **Economic diversity:** retail, wholesale, direct-to-consumer, used goods, environmental grants.
- **Ecological diversity:** global supply network of small producers.

This diversity makes Patagonia shockproof.
If one product line or market falters, others thrive, like different species balancing a forest.

Human Diversity: Designing for Difference

Diversity in teams isn't about optics; it's about pattern intelligence.
Each person represents a unique "species" of perception, sensing opportunities or risks others miss.

To build diversity that works:

1. **Hire for perspective, not just skill.** Look for people who see from different angles, disciplines, geographies, lived experiences.

2. **Design inclusive decision systems.** Ensure every voice can influence design and strategy.

3. **Measure energy, not conformity.** Teams thrive when differences are respected and energy flows freely.

A company rich in perspectives behaves like a healthy ecosystem: self-balancing and self-renewing.

Case Example: IDEO, Creativity Through Cross-Pollination

IDEO intentionally mixes disciplines on every project; designers, psychologists, engineers, anthropologists.

This integration of perspectives is what they call **"radical collaboration."**
Tension is not a problem to eliminate but a resource to manage.

Their diversity is structural, not symbolic, built into hiring, space design and workflow.
The result? Ideas that couldn't emerge from any one field alone.

Diversity doesn't slow creativity; it fuels it.

Creative Diversity: The Mycelial Network of Ideas

In a forest, the mycelial network connects roots across species, transferring nutrients and information.
In companies, creative diversity works the same way.

When ideas cross-pollinate between departments or disciplines, collective intelligence increases.
You move from linear thinking ("marketing decides") to mycelial thinking ("everyone contributes insight").

Design for creative cross-pollination:

- Pair opposites, introverts with extroverts, dreamers with analysts.
- Run idea swaps across departments.
- Use storytelling circles where everyone shares patterns they've noticed.

Every creative culture is an ecosystem waiting to be connected.

Economic Diversity: The Portfolio Ecosystem

Just as forests spread risk across many species, regenerative companies spread revenue across many flows.

Example:

- A service company adds a productized offering.
- A product company adds workshops or education.

- A local brand adds digital reach.

These secondary flows act like companion plants, protecting the core system from collapse.

Case:
Zingerman's diversified early: bakery, creamery, coffee, catering. Each arm supports others with cross-sales, shared customers and cross-learning.
Diversity became both protection and purpose.

Ecological and Supply Diversity

Ecosystems collapse when dependent on one species.
So do supply chains.

Regenerative companies build **multi-supplier systems**, relationships across scales and regions.
They value local sourcing but design redundancy in case of disruption.

This principle also applies to energy and materials: diversify toward renewables, recycled content and modular design.

Example:
Interface Carpets sources fibres from fishing nets, recycled nylon and bio-based materials, spreading risk and deepening impact.

Diversity isn't chaos; it's insurance.

The Founder's Role: From Hero to Host

In diverse systems, leadership shifts from control to curation. The founder becomes a **host**, creating the conditions for difference to collaborate productively.

Hosting diversity means:

- Inviting voices that challenge comfort.
- Designing safe structures for disagreement.
- Measuring success in connection, not consensus.

Regenerative founders don't seek uniformity.
They design choreography, rhythm, flow and trust, so difference becomes harmony.

Case Example: Dr. Bronner's, Diversity as Ethical Design

Dr. Bronner's unites unlikely groups, farmers, activists and corporate professionals, under one purpose: "We are all-one."
Their teams span continents and causes, yet cohesion thrives through shared ethics and transparent culture.

By honouring multiple identities within one mission, they prove that **unity isn't sameness.**
It's alignment through shared values.

Field Guide: Cultivating Diversity

1. Diversity Mapping

List all forms of diversity in your company:

- People (backgrounds, roles, strengths)
- Ideas (sources of innovation)
- Revenue (income streams)
- Suppliers (geographic, material)
- Partners (size, type, purpose)

Colour-code gaps. Where monoculture dominates, design new connections.

2. Edge Teams

Create temporary cross-functional teams around opportunities or problems.
Include at least one person from outside the direct domain.
Edges spark novelty.

3. Revenue Ecosystem Design

Draw your income sources like an ecosystem diagram.
Are you relying on one "species"?
Experiment with companion offerings that feed or protect your main yield.

4. Psychological Safety Rituals

Diverse teams only function with safety.
Begin meetings by setting intentions of respect and curiosity.
End by naming what was learned from disagreement.

Safety is the sunlight that lets difference grow.

5. Partner and Supply Chain Diversity

Audit your suppliers.
What percentage of spending supports small, local, women-led, or regenerative producers?
Diversify where concentration creates risk.

Case Reflection: Ecosia, Diversity as Regeneration

Ecosia integrates diversity across every dimension:

- Diverse global reforestation partners.
- Multilingual, multicultural team.
- Hybrid revenue (ads, browser extensions, partnerships).

Their diversity allows agility across climates, markets and crises.
When one region faces drought or regulation, another thrives.
That's ecosystem thinking applied to tech.

The Pattern of Diversity

In ecosystems, diversity produces three outcomes:

1. **Resilience**, more paths for recovery.

2. **Innovation**, more combinations for creativity.

3. **Stability**, more checks against imbalance.

When applied to business, diversity creates antifragility, the ability to grow stronger through disturbance.

The goal isn't maximum variety; it's functional variety, the right mix of difference to sustain vitality.

Takeaway

Uniformity feels efficient. Diversity feels messy.
But life itself is messy, and that's why it endures.

When your company honours difference as design, not decoration, you build more than inclusion; you build intelligence. Your business becomes an ecosystem, self-correcting, self-renewing and endlessly creative.

Diversity is not a value. It's a survival strategy.

The forest thrives because no leaf is the same.
So should your company.

Up Next

Chapter 12, Use Edges and Value the Marginal: The Hidden Gold of the Periphery.

We'll explore how regenerative founders uncover innovation and opportunity at the edges, in neglected ideas, underheard voices and forgotten parts of their business ecosystem.

Chapter 12, Use Edges and Value the Marginal: The Hidden Gold of the Periphery

How Regenerative Founders Find Opportunity Where Others See Waste or Risk

The Lesson of the Riverbank

Walk along a riverbank and you'll see life explode where water meets land.

Grasses root in sediment, frogs hatch in puddles, insects buzz and birds feed.

The edge, that in-between zone, is the most fertile part of any ecosystem.
It's where two worlds meet and create something new.

Ecologists call this **the edge effect**: diversity and productivity multiply where systems overlap.
The forest edge, coral reef and tidepool all prove the same rule: **life thrives in transition zones.**

In business, the same holds true.
Edges, those messy intersections between departments, industries, or ideas, are where innovation and resilience live.

The centre feels safe, but the edge creates the future.

Why Companies Ignore the Edge

Traditional organisations avoid edges because they're uncomfortable.
Edges are unpredictable, hard to measure and full of competing priorities.

So we smooth them out, standardise, streamline, centralise.
But in doing so, we sterilise.

What's left is efficient but fragile:

- Innovation stuck inside one department.
- Customers treated as data points, not collaborators.
- Communities ignored because they're "not our target

market."

Regenerative founders think differently.
They design to **live at the edges**, listening where others don't, experimenting where risk feels high and building bridges between worlds.

The Edge Effect in Organisations

Edges are where tension meets possibility.
They appear wherever two different systems interact:

Edge Type	Description	Opportunity
Internal Edge	Between teams or disciplines	Cross-pollination and innovation
External Edge	Between company and customers	Co-creation and empathy
Social Edge	Between inclusion and exclusion	Talent and insight from underrepresented voices
Market Edge	Between industries or technologies	New business models and partnerships
Temporal Edge	Between past and future	Legacy renewal and next-stage evolution

The challenge is not to avoid edges but to **harvest their energy without losing balance.**

Case Example: Zingerman's, The Edge of Community and Commerce

Zingerman's operates on the edge between a local business and a global brand.
They never franchised, yet they share their systems worldwide through training and open documentation.

This edge, local roots, global influence, became their signature. It allows them to grow without leaving home.

They also cultivate social edges: training formerly incarcerated people, supporting immigrant entrepreneurs and experimenting with open-book management.

Every edge becomes an engine of vitality.

The Pattern of Edge Thinking

In nature, edges are nonlinear and dynamic.
They require adaptability, curiosity and courage.

In business, edge thinking means:

1. **Inviting what's outside your comfort zone.**
2. **Designing collaboration between opposites.**
3. **Exploring marginal spaces for insight.**

The edge is where diversity and disturbance meet, and where learning accelerates fastest.

Case Example: Patagonia, The Edge as Ethical Frontier

Patagonia thrives by operating at cultural and market edges:

- Between business and activism.
- Between profitability and planetary responsibility.

- Between outdoor sport and political conscience.

Every time they cross an edge, they grow trust.
When they told customers, "Don't buy this jacket," they stood at the edge of contradiction and made authenticity their advantage.

Edges aren't contradictions; they're integration points.

Edges as Innovation Engines

Innovation rarely happens in the core; it happens in collisions. The best ideas emerge where two disciplines meet.

Examples:

- Psychology + Design → UX.
- Technology + Biology → Biomimicry.
- Food + Community → Social enterprise cafés.

Regenerative founders intentionally **create edge collisions** to generate novelty.

How:

- Cross-department experiments.
- Partnerships with unusual allies.
- Spaces where customers co-design solutions.

Innovation is less about genius and more about *geography*, being at the right edge.

Case Example: Dr. Bronner's, Valuing the Marginal

Dr. Bronner's sources raw materials from smallholder farmers in regions most corporations ignore, Sri Lanka, Ghana, Samoa. They turn the "marginal" into the miraculous.

Their supply chains regenerate soil, dignity and communities that global capitalism usually extracts from.

Their edge, between global brand and grassroots farming, became the soul of their business.
They didn't expand into these regions for charity.
They went there because **value lives where others don't look.**

The Marginal as Untapped Genius

What's "marginal" in your company today?

- A junior employee with ideas no one has heard.
- A product that doesn't fit the portfolio but keeps selling.
- A customer segment you can't explain but can't ignore.

These margins hold hidden intelligence.
Regenerative founders create systems that listen there.

Because what's marginal now often becomes essential later.

Case Example: Ecosia, Edges as Leverage

Ecosia turned an edge industry, internet search, into ecological restoration.
They didn't compete at the centre (Google's scale) but at the margin (ethical search).

By valuing a peripheral idea, "search funds trees", they built a loyal global following and planted over 200 million trees.

Their success came not from outpacing giants but from working at an ethical edge no one else occupied.

Edges Within Teams

Edges don't just exist between departments; they exist between people.
Between introverts and extroverts, creatives and analysts, old-timers and newcomers.

These tensions can either drain energy or generate brilliance.
The difference is **design.**

To cultivate interpersonal edges:

1. Pair opposites intentionally on projects.

2. Train teams to interpret difference as data, not conflict.

3. Celebrate friction that produces insight.

Edges aren't where things fall apart; they're where things come alive.

The Founder's Role: Edge Dweller

Regenerative founders act as **bridge builders.**
They hold space between contradictions:

- Profit and purpose.
- Growth and rest.
- Vision and pragmatism.

They don't resolve tension; they translate it.
They help the organisation metabolise difference into creativity.

This is the true leadership posture of a regenerative company: **edge literacy.**

Field Guide: Working with Edges and Margins

1. Edge Mapping

List all intersections in your company:

- Department boundaries
- Customer interfaces
- Supplier relationships
- Cultural tensions
  Mark which edges are fertile (creative) and which are eroding (conflict).
  Choose one to nourish this quarter.

2. Marginal Voices Audit

Ask: Whose voices are missing from decisions?
Bring one marginal perspective into your next planning session, a junior staff member, supplier, or customer who sees differently.
Observe how the conversation shifts.

3. Cross-Pollination Labs

Host short-term projects combining opposite skills or roles.
Give them a shared question and freedom to experiment.
Edges will reveal insights your core team can't.

4. Explore Unserved Markets

What needs or communities are "too small" for competitors to notice?
Design a micro-offering for that edge, a test product, pilot, or local service.
Edges are where niche turns to new normal.

5. Story from the Margins

Collect stories from those on the periphery of your organisation, part-time staff, contractors, old customers, suppliers.
Their experiences often hold clues to your next evolution.

Case Reflection: Interface Carpets, Learning from the Edge

Interface's journey toward zero waste began when a single customer asked an inconvenient question:

> "What's your company doing for the environment?"

At the time, sustainability wasn't on the agenda.
That "edge question" transformed the entire organisation.

Sometimes the edge enters through a single conversation, if you're willing to listen.

The Pattern of Edge Intelligence

Nature uses edges as:

- **Innovation zones**, new species evolve there.
- **Exchange zones**, energy and nutrients mix there.
- **Protection zones**, diversity buffers shocks there.

In business, edge intelligence works the same way:

- **Innovation** emerges at intersections.
- **Value** multiplies where difference meets.
- **Resilience** strengthens through connection.

The more edges you cultivate, the more adaptive your company becomes.

Takeaway

The margins hold the future.
What seems small, strange, or risky today often becomes the next system's foundation.

Regenerative founders know: evolution happens at the edges.
They listen there.
They build there.
They thrive there.

The forest's edge isn't a boundary, it's an invitation.
So is yours.

Up Next

Chapter 13, Creatively Use and Respond to Change: The Art of Adaptive Design.
We'll explore how regenerative companies treat change not as a crisis but as compost, transforming disruption into renewal through creative adaptation.

Chapter 13, Creatively Use and Respond to Change: The Art of Adaptive Design

How Regenerative Companies Turn Disruption into Fertility

The Lesson of the Forest Fire

Every few years, a healthy forest burns.
To the untrained eye, it looks like devastation.
But ecologists know better: fire is a nutrient cycle.

The blaze clears underbrush, cracks open seeds that only germinate in heat and releases minerals back into the soil.
A year later, the forest is greener, younger and more diverse.

This is the ecological truth of change:
Disturbance is not destruction. It's renewal.

Regenerative founders design their companies with that same principle.
They don't resist change; they prepare for it.
They don't just survive crises; they compost them into growth.

Because in living systems, what burns becomes soil.

The Myth of Stability

Most business education still worships stability, control, predictability, equilibrium.
But in nature, stability doesn't mean stillness.
It means dynamic balance, constant adjustment within changing conditions.

In a forest, every storm, drought, or pest outbreak reshapes the system.
Resilience isn't found in resistance; it's found in response.

When you build your company for control, it becomes brittle.
When you build for adaptation, it becomes alive.

Case Example: Patagonia, Adaptation as Integrity

Patagonia's journey is a masterclass in regenerative adaptation. They've reinvented their model repeatedly while deepening their purpose each time:

- Started as a climbing gear company → shifted to clothing.
- Began selling products → then repair, reuse and activism.
- Operated as a private company → transitioned ownership to an Earth Trust.

Each change emerged from listening; to customers, to ecosystems, to conscience.

Their principle: *"Change is not an interruption to business; it is the business."*

The Ecology of Change

In living systems, adaptation happens through three mechanisms:

1. **Sensing:** Detecting shifts in conditions.
2. **Responding:** Adjusting behaviour accordingly.
3. **Evolving:** Encoding the learning into structure.

Regenerative companies embed these mechanisms across all levels:

- **Teams** sense energy and feedback.
- **Leadership** responds through design and rhythm.
- **Culture** evolves to embody new learning.

Change becomes rhythm, not rupture.

Case Example: Zingerman's, Evolution Through Listening

Zingerman's didn't follow a growth plan; it followed an unfolding.
Every new business in its ecosystem emerged from local opportunity, often from staff ideas.

A baker wanted a better baguette → The Bakehouse was born.
A barista dreamed of roasting ethically → Zingerman's Coffee Company emerged.
Each new venture responded to real patterns, not projections.

Their growth model isn't strategy. It's **response.**

Reframing Change as Compost

Compost looks messy, it smells, steams and breaks things down.
But it's the foundation of fertility.

Every company accumulates organic matter, failed projects, missed opportunities, outdated systems.
Most throw them away.

Regenerative companies **compost them**, extracting the lessons and nutrients.

Composting Practice:

- After each failure, ask: *What did this teach us about timing, context, or capability?*
- Distil insights into new templates, rituals, or principles.
- Share compost stories openly to normalise renewal.

Change becomes learning, not loss.

Case Example: Interface Carpets, Crisis to Catalyst

When Interface's CEO Ray Anderson read *The Ecology of Commerce*, he realised his company was an ecological villain.
That shock, a moral and operational crisis, became a catalyst.

He didn't react defensively. He responded creatively:

- Reimagined the company as "a restorative enterprise."
- Set zero-waste goals decades ahead of the industry.
- Turned innovation teams loose to rethink materials.

The crisis composted into global leadership.
Interface's transformation wasn't despite change, it was *because of it.*

Designing for Adaptability

Change-resilient companies don't wait for disruption; they practice adaptation as a daily discipline.

How to design for adaptability:

1. Build Modular Systems

In nature, modular structures absorb shocks; one branch can break without killing the tree.
Design teams, products and finances with modularity:

- Independent profit centres or micro-ventures.
- Interchangeable roles and shared knowledge.
- Decentralised decision-making.

When structure is modular, change becomes local, not catastrophic.

2. Create Feedback Loops

You can't respond to what you don't sense.
Embed real-time feedback:

- Monthly pulse surveys.
- Customer story collection.
- Energy audits.
- Open financial dashboards.

Feedback turns uncertainty into intelligence.

3. Practice Safe-to-Fail Experiments

Ecosystems experiment constantly. Most seeds don't sprout, but some do.
Adopt the same rhythm:

- Run micro-experiments.
- Set clear learning goals.
- Kill what doesn't work, compost the insight and replant what does.

Small risks prevent big disasters.

4. Design Reflection Rhythms

Create structured pauses to reflect and integrate:

- End-of-cycle reviews.
- Seasonal strategy resets.
- Story-sharing rituals.

Without reflection, adaptation doesn't stick, it just repeats.

5. Protect the Core, Flex the Edge

Forests keep stable roots but change leaves every season.
Define what's *non-negotiable* (purpose, ethics) and what's *evolv-*

able (strategy, products, tools).
This gives your company both integrity and flexibility.

Adaptability without anchors becomes drift; anchors without adaptability become stagnation.

Case Example: Buurtzorg, Adaptive Autonomy

Buurtzorg's self-managing nurse teams continuously adapt to patient and community needs.
There's no manual for care delivery, just guiding principles.

When local conditions change (new regulations, patient demographics, technology), teams experiment and share solutions via a simple digital network.

Headquarters doesn't issue orders; it collects patterns.
The system evolves by learning from its edges.

This is **living governance**, change by design.

Case Example: Ecosia, Responding to Planetary Feedback

Ecosia's mission, reforestation through search revenue, is inherently tied to ecological change.
As climate and political conditions shift, so do their partnerships and strategies.

When drought reduced tree survival rates in one region, they redirected funds to restoration zones in another.
When regulation affected ad income, they diversified revenue through partnerships.

They respond to global change like a forest shifting species composition, quietly, continuously.

The Founder's Practice: Composting Chaos

Founders experience more change than anyone else in the system.
The difference between burnout and breakthrough often lies in how they interpret it.

Creative response requires three disciplines:

1. **Pause.** Let patterns surface before reacting.
2. **Reframe.** Ask, *What is this change asking of us?*
3. **Design.** Turn insight into new structure, ritual, or habit.

Crisis composted becomes clarity.

Field Guide: Practising Creative Adaptation

1. The Change Map

List major shifts (internal and external) affecting your company this year.
For each, note:

- What we can control.
- What we can influence.

- What we must accept.
  Design responses only where you have agency.

2. The Compost Journal

After any disruption, failed launch, lost client, turnover, write three reflections:

- What was the gift in disguise?
- What will we reuse?
- What will we release?

Revisit quarterly. Watch patterns emerge.

3. The Adaptive Dashboard

Track not only metrics but *sensitivity*:

- Number of experiments run.
- Time from feedback to action.
- Employee perception of responsiveness.

Adaptability can be measured.

4. Scenario Seeds

Hold quarterly scenario workshops: imagine three plausible futures (best case, crisis, unexpected).
Ask: *What structures or skills would we need to thrive in each?*

Then plant one "seed" from each scenario, a small experiment in readiness.

5. Annual Burn and Renewal

Once a year, identify one legacy process, product, or habit to intentionally retire.
Celebrate its contribution, compost its lessons and make room for renewal.
Controlled burns prevent wildfires.

Case Reflection: Basecamp, Adapting Without Losing Calm

Basecamp rebranded (from 37signals), simplified its offerings and cut entire product lines to maintain clarity.
Each change felt radical externally but was natural internally, guided by purpose, not panic.

They embody adaptive restraint: changing what must evolve while preserving what matters.
That's how adaptation remains regenerative, not reactive.

The Pattern of Creative Response

In permaculture, and in business, adaptation follows this loop:

1. **Observe**, Sense what's shifting.

2. **Pause**, Resist premature reaction.

3. **Reframe**, Translate change into opportunity.

4. **Experiment**, Take small, safe-to-fail action.

5. **Integrate**, Encode the learning into structure.

Repeat indefinitely.
This is how living systems evolve, not through planning, but through patterning.

Takeaway

Change is not a threat to life.
It *is* life.

The more your company practices creative response, the more alive it becomes.
Resilience doesn't come from resisting change; it comes from metabolising it.

The forest doesn't fear the fire.
It depends on it.

So can you.

Up Next

Chapter 14, The Final Integration: Designing a Regenerative Operating System.
In this final chapter, we'll synthesise all twelve permaculture principles into a single, living framework, a regenerative operating system for founders who want to build companies that last like forests: adaptable, abundant and alive.

Chapter 14, The Final Integration: Designing a Regenerative Operating System

How to Build a Company That Behaves Like a Living System

The Lesson of the Forest

Stand in an old-growth forest.
You'll notice there's no central manager.
No single tree leads. No department organises.

Yet the system operates with perfect coordination:
Water cycles through soil and sky.
Energy flows from sunlight to leaves to roots to fungi to seed.
Each part gives and receives in rhythm.

That is the regenerative operating system of nature, decentralised, self-regulating, adaptive and abundant.

Your company can work the same way.
The twelve principles of permaculture aren't just philosophical; together, they form a **pattern language** for designing human organisations that thrive over time.

From Principles to Patterns

Throughout this book, each principle offered a design lens:

- **Ethics:** The roots, care for Earth, care for people, fair share.
- **Principles 1–12:** The branches, leaves and canopy of practice.

But true power comes in integration, when these principles **interlock** like an ecosystem, not a checklist.

Let's see how they connect.

The Living Company Pattern

Layer	Permaculture Principle	Business Expression
Soil (Ethics)	Care for Earth, Care for People, Fair Share	Purpose, culture and governance grounded in renewal
Roots	Observe & Interact	Pattern literacy, feedback culture
Trunk	Catch & Store Energy	Asset building and knowledge systems
Branches	Obtain a Yield	Sustainable profit and multiple forms of value
Leaves	Apply Self-Regulation & Accept Feedback	Adaptive learning loops
Sap	Use and Value Renewable Resources	Human energy and ecological renewal
Canopy	Produce No Waste	Circular design and closed-loop systems
Edges	Design from Patterns to Details	Scalable coherence and elegant simplicity
Pollinators	Integrate Rather than Segregate	Interconnected teams and partnerships
Seasons	Use Small and Slow Solutions	Rhythmic, compounding growth
Biodiversity	Use and Value Diversity	Innovation through difference
Ecotone	Use Edges & Value the Marginal	Creative tension and inclusion

When integrated, these become your **Regenerative Operating System (ROS)**, a model that regenerates resources, relationships and results over time.

The Regenerative Operating System (ROS)

A living business doesn't just *use* these principles; it *runs* on them.
Here's how your ROS functions in practice:

1. Ethics as Operating Boundaries

Everything begins and ends with the three ethics:

- *Care for Earth* = ecological responsibility
- *Care for People* = human sustainability
- *Fair Share* = circular distribution of yield

These aren't slogans; they're **filters for every decision.**
If a choice violates one ethic, it depletes your soil.

2. Feedback as Nervous System

Every living system feels its environment and adjusts.
Your feedback loops, customer data, staff check-ins, energy audits, are the sensory network of your company.
The tighter the feedback, the faster the learning.

3. Knowledge as Memory

Regenerative companies store experience like forests store carbon, in living, reusable form.
Your documentation, debriefs and reflection rituals are your company's **soil memory.**
They make wisdom renewable.

4. Energy as Circulation

Money, trust and creativity are your sap.
When they circulate freely, the company stays alive.
When they stagnate or leak, decay sets in.

Design flow systems, recurring revenue, transparent communication, equitable profit sharing, to keep energy moving.

5. Diversity as Resilience

Monocultures may grow fast but collapse quickly.
Diversity in people, products and perspectives creates self-healing capacity.
When one element falters, others adapt.

Diversity is your insurance policy and your source of evolution.

6. Small and Slow as Rhythm

The forest doesn't rush growth; it layers it.
Your company grows strongest when its rhythm matches its capacity.
Use cycles, short bursts, long rests, to let learning sink in.

The faster you pause, the faster you evolve.

7. Edges as Innovation Zones

Your most fertile ideas will come from boundaries — between teams, customers, or contradictions.
Design to keep edges alive: cross-functional projects, listening

posts, partnerships with the unexpected.
Innovation lives in the friction.

8. Waste as Resource

Every problem hides potential.
Turn unused materials, time and lessons into inputs for new processes.
Your ability to **compost mistakes** determines your fertility.

9. Change as Pulse

Instead of chasing stability, train your company to move with change.
Forecast less, sense more.
Respond quickly, reflect deeply, evolve constantly.

Change is not a threat to your business model; it's your metabolism.

Case Example: Zingerman's as a Living System

Zingerman's entire community of businesses operates as a regenerative organism:

- **Ethics:** Shared purpose and profit distribution.
- **Feedback:** Open-book management and transparent dialogue.
- **Knowledge:** Training programs that store collective wisdom.

- **Diversity:** Dozens of ventures connected by common DNA.
- **Edges:** Collaboration between food, education and consulting.
- **Slow Growth:** 40 years of compounding resilience.

The result is an organisation that **keeps becoming itself**, evolving without losing identity.

Case Example: Dr. Bronner's as Ethical Infrastructure

Dr. Bronner's demonstrates how a company can turn ethics into architecture:

- Supply chains regenerate ecosystems.
- Employees share profits and governance.
- Surplus circulates into activism and global partnerships.

Their ROS isn't a strategy document; it's culture in motion. Each principle is expressed in policy, ritual, or relationship.

The Four Capacities of a Regenerative Company

A regenerative company isn't defined by size or industry, but by **capacity.**
When your systems are healthy, these four capacities emerge naturally:

1. **Awareness**, the ability to sense change and feedback.

2. **Adaptation**, the ability to respond and redesign.

3. **Circulation**, the ability to keep energy moving and multiplying.

4. **Regeneration**, the ability to renew resources faster than you use them.

These capacities become your **core KPIs**, not just profit, but fertility.

Building Your Regenerative Operating System

To implement your ROS, follow these design steps:

Step 1: Map Your Ecosystem

Identify all stakeholders, staff, customers, suppliers, community, planet.
Draw energy flows between them: money, trust, communication, waste, ideas.
You'll see where feedback loops break or energy leaks.

Step 2: Anchor in Ethics

Write your company's three ethical commitments in plain language.
Example:

- "We only grow in ways that replenish."
- "We design work that sustains human life, not drains it."

- "We share surplus to strengthen the whole."
  Use these as filters for all decisions.

Step 3: Layer the Principles

Design initiatives that embody multiple principles at once. Example: a circular product (no waste) made by a self-managing team (integration) with renewable materials (resource value). Stacking principles increases regenerative yield.

Step 4: Create Feedback & Reflection Rhythms

Schedule cycles for reflection, weekly team check-ins, quarterly compost sessions, annual system reviews.
Regeneration requires rhythm.

Step 5: Reinvest Surplus into Renewal

When profit, time, or energy accumulate, reinvest them in soil:

- Education and rest for people.
- Infrastructure for sustainability.
- Partnerships for ecological or social repair.

Surplus only becomes wealth when it circulates.

Case Reflection: Ecosia, The Digital Forest

Ecosia operates as a living ecosystem:

- **Root:** Clear ethics, reforestation, privacy, transparency.

- **Flow:** Revenue circulates directly into tree planting.
- **Feedback:** Regional monitoring and adaptation.
- **Adaptation:** Constant reallocation based on climate shifts.
- **Regeneration:** Trees restore soil, community and carbon.

They've built a regenerative operating system at a planetary scale, a model for the next economy.

The Pattern of Regenerative Maturity

When companies integrate all principles, they move through these stages:

1. **Extractive:** Growth through depletion.
2. **Sustainable:** Growth with offsetting harm.
3. **Regenerative:** Growth that heals the systems it depends on.
4. **Evolutionary:** Growth that regenerates other systems too, communities, industries, ecosystems.

Your company's purpose expands as its maturity deepens.

The Regenerative Founder's Role

You are not a CEO in the traditional sense.
You are a **gardener of systems.**

Your daily work is to:

- Observe patterns.
- Design flows.
- Tend culture.
- Protect rhythm.
- Compost failure.
- Harvest insight.
- Circulate surplus.
- Replant purpose.

You don't drive the company. You steward its ecosystem.
And in doing so, you become part of a larger cycle, the regeneration of economy itself.

Field Guide: Your Regenerative Operating System

1. Ethics Review

Ask before every major decision:

- Does this replenish or deplete?
- Who benefits and who bears the cost?
- How does this choice serve future generations?

2. Flow Audit

Track money, trust, time and knowledge.
Are they circulating or pooling?
Where flow stops, life stops.

3. Compost Session

Hold quarterly "compost reviews."
List what's outdated, redundant, or draining.
Retire it, learn from it and reinvest the energy elsewhere.

4. Seasonal Planning

Design your year in regenerative rhythm:

- **Winter:** Reflection and compost.
- **Spring:** Experimentation.
- **Summer:** Production and yield.
- **Autumn:** Harvest and sharing.

This rhythm sustains energy through seasons of life and business.

5. Yield Mapping

List all yields your company generates, financial, social, human, ecological, cultural.
Ask how each can reinforce another.
Build loops of reciprocity.

The Regenerative Equation

Regeneration = Ethics × Design × Feedback × Time

Ethics provide direction.
Design creates structure.
Feedback ensures flow.
Time compounds fertility.

Sustainability maintains.
Regeneration multiplies.

Takeaway

A permaculture company isn't a metaphor.
It's a living design, an organisation that grows in relationship with the world around it.

When you integrate the twelve principles, your company stops being a machine that extracts value and becomes an ecosystem that **creates it.**

Your culture becomes soil.
Your people become stewards.
Your profit becomes seed.

The forest doesn't need permission to grow.
Neither do you.

What Next?

Epilogue: Beyond Hustle

We began this journey asking:
How do we build companies that thrive without burnout, exploitation, or endless growth?

The answer isn't more hustle. It's better design.
Design rooted in life's patterns, circular, adaptive, relational, slow.

To build a regenerative company is to remember what all living systems know:
Nothing in nature hustles.
Everything in nature belongs.

When your business learns to belong, to place, to people, to purpose, it will never run out of energy again.

Acknowledgements

This book is built on the work of thousands of people who have tended the soil of regenerative thinking.

To **Bill Mollison and David Holmgren**, who gave us permaculture; the design science that inspired this book.

To the **farmers, activists and entrepreneurs** who translated those principles into food systems, communities and companies.

To **Zingerman's, Patagonia, Dr. Bronner's, Interface, Ecosia, Buurtzorg and Basecamp**, each living proof that regeneration and profit can coexist.

To the **readers and builders** who kept asking: "What would it look like if business worked like nature?"
This book is my answer... and my thank you.

To my wife, Skie, my kids, Tamar, Zoe & Asher and my extended family, John, Donna, Binny andy & the rest of the tribe, thanks for showing me what real connection looks like.

And to **the next generation of founders**, who will inherit both the damage and the possibility of this era.
May you build differently.
May you build for life.

Closing Reflection, The Long View

One day, your company will outgrow you.
If you've built regeneratively, that will be your greatest success.

A regenerative company is like a forest.
You plant the seeds. You tend the soil. You compost the failures.
Then, someday, you step back and realise it can now sustain itself, feeding others, sheltering life, creating beauty long after you've gone.

That's what it means to build something that lasts.
Not a monument, but a meadow.
Not an empire, but an ecosystem.

The world doesn't need more founders who hustle harder.
It needs founders who design with life.
Founders who build organisations that breathe, rest and renew.

That's the work now.
To build companies that, like forests, give more than they take.

If you remember only one thing from this book, let it be this:
You are not here to extract. You are here to participate.
In the great pattern of regeneration.

And the moment you choose that pattern, your business... and perhaps your life... begins to grow in a new direction.

Made in the USA
Monee, IL
04 November 2026

67380858R00098